Molecular Simulation Studies in Material and Biological Sciences

MOLECULAR SIMULATION STUDIES IN MATERIAL AND BIOLOGICAL SCIENCES

KHOLMIRZO KHOLMURODOV
EDITOR

Nova Science Publishers, Inc.
New York

For permission to use material from this book please contact us:
Telephone 631-231-7269; Fax 631-231-8175
Web Site: http://www.novapublishers.com

Library of Congress Cataloging-in-Publication Data
International workshop : molecular simulation studies in material and biological sciences / Kholmirzo Kholmurodov, editor.
p. cm.
Includes index.
ISBN 13 978-1-59454-607-5
ISBN 10 1-59454-607-X
1. Biology--Computer simulation--Congresses. 2. Materials science--Computer simulation--Congresses. I. Kholmurodov, Kholmirzo.
QH324.2.I559 2006
572.801'13--dc22 2006002075

Published by Nova Science Publishers, Inc. ✦New York

CONTENTS

PREFACE

The methods of computer molecular simulations of large molecular systems, first proposed more than 50 years ago with the invention of computers, have shown an outburst of development in the last 5-10 years. With the creation of new parallel vector supercomputers and special purpose computational clusters, the molecular simulation methods are becoming a real tool in bioengineering, nanotechnology and material science, capable of estimating the details of the atomic scale processes and other technologically interesting phenomena. The molecular simulation serves as a practical tool in the development of new materials and new drugs, performing the large scale calculations on molecular complexes of hundreds, thousands or even multimillion particle systems. The molecular simulation techniques allow, for example, to directly explore the conformational changes and dynamical processes of DNA and proteins, to estimate the phase transition behavior of polymeric systems, to study the cluster-surface nanostructures in physical chemistry, biochemistry, and many others.

Chapter 2 discusses the application of linear-scaling semiempirical quantum mechanical methods to molecular biophysics. Three problems have been selected:

1. the computation of the macrodipole of α-chymotrypsin in water,
2. the prediction of the protonation state of Sialyl Lewis X tetrasaccharide when interacting with the protein E-selectin, and
3. the interaction of the SH2 domain of protein kinases, a key element of intracellular signal transduction events, with a phospho-peptide.

Molecular dynamics (MD) simulations using the Collision Dynamics method are described in Chapter 3, for several model membranes:

i. hydrated homogeneous lipid bi-layers constructed of saturated and unsaturated phosphatidylcholines (PC), – 1-stearoyl-2-oleoyl-*sn*-glycero-3-PC, 1-stearoyl-2-linoleoyl-*sn*-glycero-3-PC, 1-stearoyl-2-linolenoyl-*sn*-glycero-3-PC, 1-stearoyl-2-arachidonoyl-*sn*-glycero-3-PC, 1-stearoyl-2-docosahexaenoyl-*sn*glycero-3-PC;
ii. the above-listed PC bi-layers with a cylinder hole and the PC bi-layers with a peptide inclusion;
iii. the PC bi-layers with several sphere-cylindrical holes and the PC bi-layers with a low mole fraction of cholesterol.

Several results of the investigations are discussed for the homogeneous bi-layers. The $C{-}H$ and $C{-}C$ bond order parameter $-S_{CH}$, $-S_{CC}$ profiles of the hydrocarbon tails, the bond orientation distribution functions and the root-mean-square values of the positional fluctuations of the lipid chain carbons, mass density profiles along the bi-layer normal (Z) for different atoms and groups of atoms were calculated. In the homogeneous bi-layers the order parameter characteristics of lipid hydrocarbon chains in the "liquid" region of the bi-layer (somewhat remote from the membrane-water interface) are found to be qualitatively similar to those of single unperturbed hydrocarbon chains: the behavior of the acyl chains in this region is dominated by the intramolecular short range interactions. A close relationship between the physical characteristics and the structure of the chains is elucidated: the number of chain carbons, the number of cis double bonds and their position in the chain determine principally the calculated properties.

In Chapter 4, some of the structural and functional processes taking place in the mono- and multi-lipid bi-layers are investigated at time scales up to 10 ns using the methods of computer simulations. We report on the development of a special code "MDesigner" intended for construction of lipids and lipid bi-layers.

Molecular dynamics (MD) simulations were performed for six model monolayers of diacylglycerolipids (DG) with a different number of cis double $C = C$ bonds of the sn-2 hydrocarbon chains. The simulation monolayers consisted of 48 identical lipid molecules arranged in rectangular simulation cells, with periodic boundary conditions in the surface plane (X, Y directions). The temperature (T), the number of atoms (N) and the surface area per molecule (A) were fixed; two series of the MD simulations were performed under two different (fixed) temperatures. The simulation parameters have provided a liquid crystalline state of the monolayers, the MD trajectories were 5000 ps. The effect of the degree of unsaturation of the sn-2 hydrocarbon chains as well as the temperature influence on the lipid properties have been studied in Chapter 5. The results of the simulations are compared with the available experimental data and other simulations of similar systems.

Chapter 6 reviews the recent development of the algorithm for linear response functions of quantum systems described by a very large Hamiltonian matrix. These numerical methods are combinations of a numerical solution of the time-dependent Schroedinger equation, random vector representation of trace, and Chebyshev polynomial expansion of Fermiand Boltzmann-operators, whose application includes optical absorption of nanocrystallites, and electron spin resonance (ESR) spectrum of one dimensional antiferromagnets at finite temperature.

One of the difficulties in studying the internal structures of the earth and other planets comes from the fact that there are no direct methods for observing the inside of planets. For decades the technique of laboratory experiments at high pressure and high temperature equivalent to those inside planets has been rapidly developing and providing new information that is unavailable by observations only. The first principle molecular dynamics calculation is recently attracting astrophysicists' attention as the third method, which is complementary to indirect observation and laboratory experiment. In Chapter 7 we review our latest study of the first principle of molecular dynamics focused on the two newly discovered crystal structures, $MgSiO_3$ post-perovskite and methane (CH_4) hydrate phase III.

Chapter 8 presents Molecular Dynamics (MD) simulations and spectral studies, which were performed on visual pigment rhodopsin to investigate its dark-adapted and transition states, intermediates and conformation properties. Based on the MD simulations with a supercomputer and special-purposes MDGRAPE-2, machine we investigated for the rhodopsin molecule, the structural conformational changes and dynamical behaviors. We performed 3 ns MD-calculations on the rhodopsin and presented the structure analysis data for its dark-adapted state. We have compared the dynamical properties of the rhodopsin with regard to the following considerations: (1) Rhodopsin without chromophore retinal; (2) Rhodopsin with chromophore (11-*cis*) retinal. In the first approach for the retinal chromophore all interactions inside the rhodopsin pocket, as well as with the water solvent, were neglected. The RMSD (root-mean-square deviation) simulation data for the helices I-VII of rhodopsin (without retinal) were performed. From the analysis of the RMSD results we observed a relative movement of rhodopsin helices, which possesses different behavior between the helices. The structure images of rhodopsin were constructed and the representative configurations were compared for the 3 ns state. The results showed the largest displacements from the reference structure for the helices III and V. In the second consideration we have fulfilled the RMSD and structural analysis for the rhodopsin (with 11-*cis* retinal), generated the pictures of the atomic-scale processes for the binding pocket, surrounding the chromophore retinal, and compared the helical deviations for the beta-ionone ring and Schiff base linkage regions of the protein. We have observed that the inclusion of a small retinal chain into the whole rhodopsin dynamics (+ solvent) does not influence the RMSD behavior of the total rhodopsin structure. Within the 3-ns calculations, the RMSD of the rhodopsin (with 11-*cis* retinal) possesses the same average value and behavior as in the absence of the retinal chromophore. However, it should be stressed that with regard to selected rhodopsin helices the inclusion of the retinal chromophore changes all RMSD values and causes essential helical distortions.

The most remarkable point of our observations is that the rhodopsin helical distortions in the dark state are accompanied with the transformation of the retinal chromophore, viz. with the rotation of the beta-ionone ring inside the protein binding pocket. From our simulation data we observe that within a short time [0.3-0.4] ns, the beta-ionone ring changes its orientation in relation to the polyene chain (chromophore retinal of 11-*cis* conformation). The beta-ionone ring at later stages of the dynamics takes a more planar position with regard to the helix III of the rhodopsin. Within this time interval we observe that the amino acid residue Phe-261 changes its orientation relative to the retinal chromophore (in beta-ionone region) as well. The positions of the two key amino acid residues Phe-212 and Ala-269 change with the beta-ionone ring rotation. Starting from this interval [0.3-0.4] ns the residue Ala-269 approaches closer to the retinal site (beta-ionone part) than the residue Phe-212.

The low-temperature absorption spectroscopy technique has been used to study the primary stages of rhodopsin photolysis. The digitonin extract of bovine rhodopsin was irradiated at -155 _C by blue light (436 nm). The following dark spectral changes have been recorded in the course of a slow temperature rise. The simultaneous appearance of more than one spectral product has been revealed throughout bato- to lumirhodopsin transition in the temperature interval from -155 _C to -145 _C. An intermediate product which is transformed along with batorhodopsin to the next product and known as a blue-shifted intermediate [1] was observed. The structural transformation properties of rhodopsin were discussed in due course of the stimulations and experimental observations. It is proposed that the appearance of more than one intermediate at each stage of rhodopsin photolysis reflects the existence of multiple conformation states of the retinal chromophore in rhodopsin molecule in the dark-adapted state. As a summary of our observations we suggest that the conformation dynamics of the opsin helices determines the most probable chromophore (11-*cis* retinal) configuration, so that a favorable retinal conformation has to be adjusted by the protein surroundings.

Experimental and theoretical aspects of the study of fullerene clusters in molecular and colloidal solutions are presented in Chapter 9. Experimental data obtained by different techniques, including small-angle neutron scattering from a number of solutions of fullerenes, are summarized covering molecular solutions of C_{60} fullerene in carbon disulfide, colloidal solutions of C_{60} fullerene in water, nitrogenated solvents and binary liquids. Conditions and mechanisms of cluster formation in these solutions are considered. Applications of molecular dynamics simulations for modeling the considered solutions are discussed.

The features of the fullerene-solvent interaction in the system fullerene C60/carbon disulfide are analyzed by means of the molecular dynamic simulations. A specific organization of the solvent molecules close to the fullerene surface is revealed. It can affect significantly the visible size of fullerene in this solvent determined by different techniques. The origin of such organization is discussed in Chapter 10.

As explained in Chapter 11, Molecular Dynamics (MD) simulations have been performed on the parallel-vector and special-purposes MD machines to study the dynamic behavior of a flexible polyethylene (PE) chain subjected to various heating conditions. Both the effect of the short-range and electrostatics interactions on the dynamics of the PE system was estimated and compared. The main objective of the present study was to investigate the influence of the heating condition on a final state of the polymeric system and sensitivity of its thermodynamic quantities (density, energy, etc.) on the heating modes. For this purpose the three different simulations were performed regarding the following regimes: fast, moderate, and slow heating processes. The peculiarities of the main thermodynamic quantities of the system were estimated in detail and compared between all three processes. Molecular, mechanical and MD calculations on the PE system were performed using the optimized version of the DL POLY code, adapted for the vector- and special-purpose MDGRAPE-2 machines.

In Chapter 12 we describe an object-oriented C++ implementation of the three-dimensional Fast Multipole Method (FMM) for evaluation of the Coulombic potential and the force field for large systems of charged or gravitating particles. The FMM is the preferable alternative for the evaluation of the Coulombic forces in the molecular dynamic simulations of large molecules. In contrast to the direct summation $\sum_{j\neq i}\frac{1}{r_{ij}}$, that scales as $O(N^2)$ with the number of particles, the FMM method has $O(N \log N)$ asymptotic. The FMM algorithm is extremely important for simulation of large biomolecules, first of all for proteins and nucleic acids, and for the simulation of large gravitating systems is astrophysics. The implementation has been performed according to the Greengard- Rokhlin algorithm [*Acta Numerica*, 1997]. The code was tested on an ordinary PC for a systems of up to 10^6 particles.

As outlined in Chapter 13, the central role that cyclin-dependent kinases play in the timing of cell division and the high incidence of genetic alteration of CDKs or deregulation of CDK inhibitors in a number of cancers make CDC28 of yeast *Saccharomyces cerevisiae* a very attractive model for studies of mechanisms of CDK regulation. We have found that certain gene mutations, including *cdc28-srm,* affect cell cycle progression, maintenance of different genetic structures [1] and increased cell sensitivity to ionizing radiation [2]. A *cdc28- srm* mutation is not a temperature-sensitive mutation and differs from known *cdc28-ts* mutations because it has the evident phenotypic manifestations at 30°C. Sequencing analysis of *cdc28-srm* revealed a single nucleotide substitution G20S. This is a third glycine in a conserved sequence GxGxxG in the G-rich loop positioned opposite the activation T-loop. Despite its demonstrated importance, the role of the G-loop has remained unclear. The crystal structure of the human CDK2 has served as a model for the catalytic core of other CDKs, including CDC28. Nanoseconds long molecular dynamics (MD) trajectories of the CDK2/ATP complex were analyzed. The MD simulations of substitution CDK2-G16S (CDC28-G20S) shows a conformational change of CDK2 structure resulting in the movement of the G-loop away from ATP and a new rearrangement of amino acids in the T-loop.

In Chapter 14, we used molecular dynamics (MD) simulations to study the density and the temperature behavior of a flexible polyethylene (PE) subjected to various heating conditions, and to investigate the PE chain conformational changes in a water solvent. First, we have considered the influence of the heating process on the final state of the polymeric system and the sensitivity of its thermodynamic characteristics (density, energy, etc.) for different heating regimes. For this purpose the three different simulations were performed: fast, moderate, and slow heating. Second, we have investigated the PE chain conformational dynamics in water solvent for various simulation conditions and various configurations of the environment. From the obtained results we have the pictures of the PE dynamical motions in water. We have observed a protein-like behavior of the chain PE, like that of the DNA and the proteins in water, and have also estimated the rates of the conformational changes. For the MD simulations we used the optimized general-purpose DL POLY code (B. Lustig, Nucleic Acids Research, 1995; G. Lancelot, Proc. Nat. Acad. Sci., 1997) and the generic DREIDING force field. The MD simulations were performed on the parallel computers and the special-purpose MDGRAPE-2 machine.

Simulation of distribution of elastic perturbation along a linear polyethylene chain in polymeric crystallite was carried out in Chapter 15. Like that in Frenkel-Contormodel, influence of crystal substrate was taken into account by introduction of sine intermolecular potential. For the description of interatomic interaction along the molecular chain the Morse potential was used. In the system of N =100 methylene groups with and without structural defects in a chain, oxidized groupings CO with a variation of their number (1 and 2 groupings), the place of localization and interposition were considered. The study was carried out by the method of molecular dynamics (MMD). Elastic perturbation was transferred along molecular chain by a stretching of trailer bond, the dynamics of which was traced in a time interval. Abnormal character of this perturbation in places of localization of defects were shown.

In: Molecular Simulation Studies in Material ... ISBN 1-59454-607-X
Editor: Kholmirzo Kholmurodov, pp. 1-2

Chapter 1

MOLECULAR DYNAMIC SIMULATION OF BIO- AND NANOSTRUCTURES

Konstantin Shaitan *
M. V. Lomonosov Moscow State University
119899 Khokhlov st. 1-12, Moscow, Russia

Molecular dynamics simulations are widely used to study structure and functions of macromolecules for more than 30 years. The treatment of the classical motion equations (Newton equations) for the system of interacting atoms is the basis of this method. The force field usually can be defined as a sum of partial interatomic interactions. If the force field reflects real interactions, and numerical algorithms are reliable, the treatment of some dozens or hundred thousands of equations describes dynamical behaviour of the macromolecular system with such details as it impossible to reach in the experiment.

There are two main approaches to MD simulations. First, equilibrium MD simulations or MD performed under interatomic interactions only. Second, steered MD (SMD) or MD under interatomic interactions and special external forces applied to given atoms.

To compare MD simulations and experimental results we should satisfy the following aminoacid residues conditions:

- energy uniform distribution over the molecular degrees of freedom should be performed;
- trajectories should be ergodic (that is the representative point should scan all significant regions of the configuration space); this condition is hardly possible for macromolecular systems.

To compare SMD and experimental results it is important that:

- energy uniform distribution over the molecular degrees of freedom should be performed except separate ones; and
- some additional conditions that depend on peculiarities of given system should be met.

*E-mail address: shaitan@moldyn.ru

The examples of the systems that have recently been studied in our lab by SMD:

- dynamics of the biomembrane piercing by the nanotube;
- complex of polyalanin with nanotube;
- nanosyrindge model (some sort of artificial virus);
- polymer chain refolding by pulling through model nanotube under special force field design.

The illustrative picture of molecular system dynamic behaviour can be obtained comparatively fast using well known techniques. However, calculation of the real physical picture of polymer and biopolymer systems is not easy and it requires significant theoretical efforts.

The small peptides equilibrium dynamics are discussed as well. The dynamics properties of even simple peptides are not sufficiently clear. In particular we may think about the peculiarities of dynamics of natural aminoacid residues in comparison to their homologous and isomers. There are about 20 natural aminoacid residues marked somehow in the range of the similar molecules. How could it be explained that some aminoacid residues sequences have funnel on their energy landscape that provides folding. The detailed and comparative analysis of oligopeptides conformational dynamics is discussed. We developed special technique by using Poincare cross sections, auto- and crosscorrelation functions and cluster analysis. It is very important that for such small peptides we could use the trajectories with good ergodic properties. An MD simulation for the series of 4 hundred dipeptides and 20 monopeptides was done using special protocol. This protocol enables us to scan by representative point all available configuration space of the system. Trajectory length was 10 ns. For acceleration the potential barriers overcoming the effective temperature of 2 thousand Kelvin degrees, as well as collisional thermostat, were used. We would like to stress that the equilibrium energy distribution on the molecular degrees of freedom was under control. Peptide dynamics in the virtual, water and methanol media was studied.

Some of the results are quite surprising. The set of torsion angles of autocorrelation functions that represent the individual dynamic behaviour of the degree of freedom is separated into relatively small number of discrete groups. Inside the group the auto correlation functions are in good agreement. That is there is dynamic similarity of torsion angels inside each group. These groups do not correspond to the types of torsion angels φ, ψ or χ. Torsion angels of different types can be in one group. On the other hand, the angels of one type can be in different groups. We suppose that it is not an accident. It suggests a definite ordering or symmetry of energy landscape of natural peptides. It may be significant for funnel appearance. Natural aminoacid residues are very similar in their dynamics in general. Angles φ and χ dynamics are more similar. The angles ψ dynamics are more variable. Residues dynamics depend on their positions. Sometimes the position is more important than the type of residue.

When coming to homologous and isomers of natural aminoacids, the dynamics changes are very significant. Even changing in position of OH group in Tyr ring leads to strong variations in dynamics and free energy maps. Water medium smoothes the free energy landscape of the peptides mentioned above.

In: Molecular Simulation Studies in Material ... ISBN 1-59454-607-X
Editor: Kholmirzo Kholmurodov, pp. 3-13

Chapter 2

QUANTUM BIOPHYSICS WITH LINEAR-SCALING ELECTRONIC STRUCTURE METHODS

Fabio Pichierri*
COE Laboratory of Computational Chemistry
Tohoku University, 2-1-1 Katahira,
980-8577 Sendai, Japan

Abstract

This chapter discusses about the application of linear-scaling semiempirical quantum mechanical methods to molecular biophysics. Three problems have been selected:

1. the computation of the macrodipole of α-chymotrypsin in water,
2. the prediction of the protonation state of Sialyl Lewis X tetrasaccharide when interacting with the protein E-selectin, and
3. the interaction of the SH2 domain of protein kinases, a key element of intracellular signal transduction events, with a phospho-peptide.

1. Introduction

Electrons play a crucial role in important biophysical processes such as electron transfer, enzymatic reactions, photophysical processes, etc. It follows that the knowledge of the wavefunction (ψ) of proteins and other biomolecules is a necessary condition for approaching such complex problems [1]. So far, however, the application of quantum mechanics to proteins has been hampered by the very large number of atoms ($> 10^3$) that characterize such systems. As a result, proteins have been traditionally investigated with the aid of force-field methods, namely molecular mechanics (MM) and molecular dynamics (MD) simulation, whereas the application of quantum chemistry methods has been limited to smaller molecules [2].

In order to solve the Schrödinger equation for systems containing thousands of atoms some approximations are necessary. In this regard, semiempirical quantum mechanical

*E-mail address: fabio@tagen.tohoku.ac.jp

methods are well suited to approach systems that are considered large from the chemist's viewpoint [3]. With the use of canonical molecular orbitals (MOs) it is generally possible to study molecules containing a few hundred atoms. In the 1990s, however, Stewart had introduced an original linear-scaling algorithm which allows the application of semiempirical quantum mechanics to systems containing thousand of atoms [4]. In the next paragraphs we will describe the key features of this approach and provide three examples of applications in molecular biophysics.

2. Computational Methods

Semiempirical quantum mechanical methods were developed back in the 1970s with the aim of providing an approximate tool for studying the electronic structure of chemically relevant systems that were intractable with more accurate MO-LCAO methods. The most popular semiempirical methods are those based on the Neglect of Diatomic Differential Overlap (NDDO) approximation [3], which neglects the overlap among orbitals residing on different atomic centers. The consequence of this approximation is that the corresponding interaction integrals are zeroed. The main NDDO parameterizations are Dewar's MNDO (Modified of Neglect Differential Method) and AM1 (Austin Model 1) methods, and Stewart's PM3 and PM5 (Parameterized Method number three and five) methods [5]-[7]. Molecules with a few hundred atoms can be approached with such methods. However, up until recently larger systems were still problematic to treat even on powerful workstations.

In 1996 Stewart has published a seminal paper describing an original linear-scaling approach that, in conjunction with semiempirical NDDO methods, allows one to investigate systems such as proteins and enzymes comprising several thousand atoms [4]. The code was originally named MOZYME and is now part of the MOPAC2002 software package [8]. His method employs localized molecular orbitals (LMOs) which, unlike canonical MOs, are localized over few atoms. Figure 1 compares the conventional SCF procedure with the LMO-based procedure invented by Stewart [4]. The starting LMO set, which is necessary to construct the initial density matrix, is obtained from the Lewis diagram of a closed-shell molecule. Simply put, the Lewis diagram represents the valence electronic structure of a molecule in terms of single bonds, π-bonds, lone-pairs, etc. The LMOs are required to be orthonormal. Optimal LMOs should involve one or two atoms at most. The cycle is repeated over and over until the self-consistent field (SCF) is achieved. At this time of writing, geometry optimizations with the SCF-LMO method are possible in the gas-phase only. Solvent effects can be modeled with the linear-scaling version of the Conductor-like Screening Model (COSMO) of Klamt [9],[10]. At present, only single-point LS-COSMO calculations are possible. When these methods are employed to study the electronic structure of proteins, it is recommended to first relax their experimental geometries with force-field methods (MM or MD). Subsequently, single-point (1SCF) calculations with the LS-COSMO model are being performed. If crystal structures have been determined at atomic resolution (i.e. beyond 1 Å), then the energy minimization step can be skipped, provided that hydrogen atoms are already present in the original crystal data. Care must be exercised in checking possible missing residues and in reproducing the correct protonation state of the protein at a given pH. In what follows we will show how linear-scaling semiemiempirical quantum mechanical (LS-SQM) methods can be successfully applied to investigate complex problems in the fields of biology and biophysics [11]-[13].

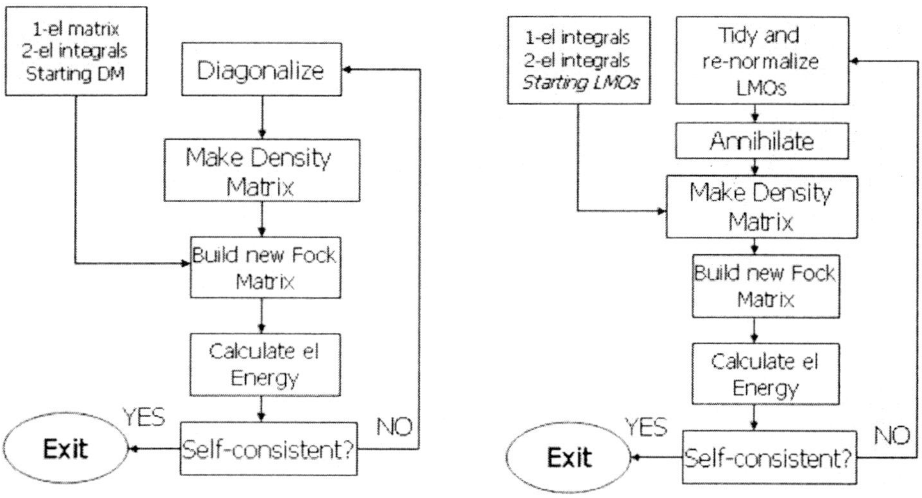

Figure 1. Comparison between the (left) canonical SCF procedure and (right) the LMO approach. Modified from reference [4]

3. Applications

3.1. α-chymotrypsin Macrodipole

Proteins contain lot of charged amino-acid residues and, hence, their permanent dipole moments can reach magnitudes of several hundred Debyes. For this reason they are called macrodipoles [14]. So far, and to the best of our knowledge, macrodipoles have been computed with the aid of empirical methods but not yet from quantum mechanical principles. Given the availability of LS-SQM methods, we decided to calculate the macrodipole of a protein from its wavefunction. The protein selected for this purpose was α-chymotrypsin (αCT) whose 3D structure is shown in Figure 2. αCT is a serine protease that cleaves the peptide bonds of polypeptides. The active site of αCT is made of three key residues, namely an aspartic acid residue (Asp102), an histidine residue (His57), and a serine residue (Ser195), respecetively, as shown in Figure 3. These three residues, called the catalytic triad, are located right in between the antiparallel β-domains, each of which is provided with six β-strands (see Figure 2). The catalytic triad is activated at above pH 7, upon deprotonation of Asp102. In the enzymatically active form, the catalytic triad employs the hydroxyl group of its Ser195 residue to carry out the nucleophilic attack on a nearby peptide bond which is part of a polypeptide interacting with αCT.

For our computational study [11] we chose the crystal structure of bovine αCT determined by Tsukada and Blow at 1.68 Åresolution [15]. The atomic coordinates of this structure (A chain), however, cannot be used directly for the QM calculation since H atoms need to be added. The addition of H atoms was performed by reproducing the protonation state of the protein at pH 7. The choice of this pH value was dictated by the fact that at this pH the catalytic triad is well characterized and the dipole moment of the protein has been measured. The resulting structure was energy-minimized with the consistent force field -

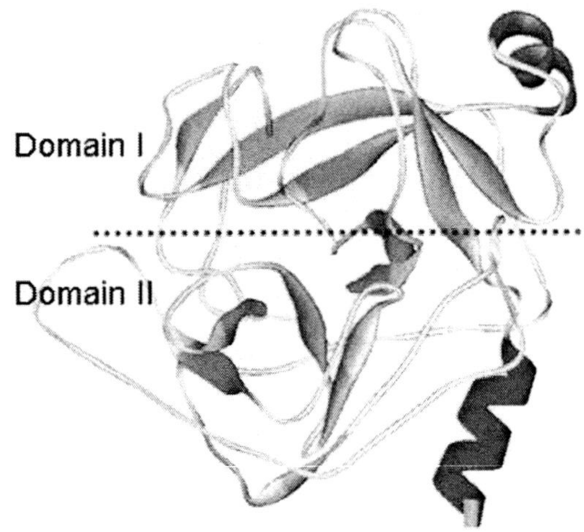

Figure 2. 3D structure of the protein α-chymotrypsin

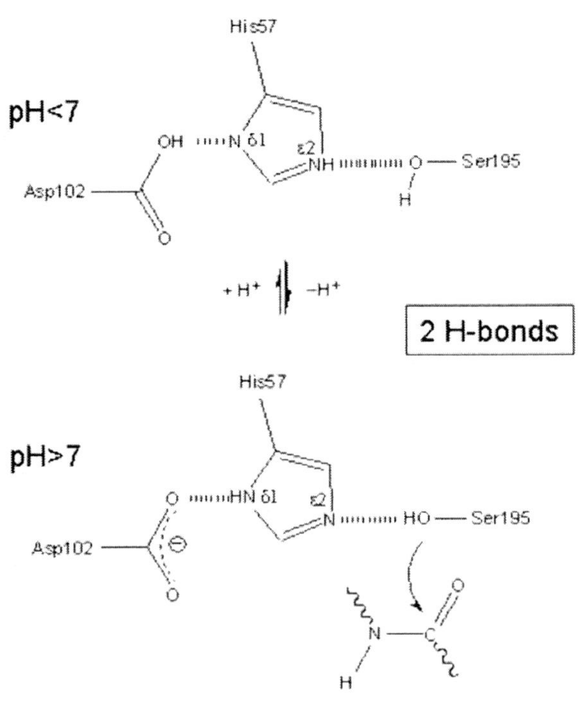

Figure 3. The catalytic triad of α-chymotrypsin

molecular mechanics method (CFF-MM) and by employing a distant-dependent dielectric constant recommended for water. The RMSD of the relaxed model (backbone atoms only) corresponds to 0.78 Å. This structural model, which comprises 3493 atoms, was employed in the subsequent QM calculations. At pH 7 the Lewis structure of the solvent-free protein has net charge 3+.

An initial set of QM calculations were performed in the gas-phase. Proteins, however, operate in water and, hence, solvation effects must be taken into account. These have been included in two ways. The first is through the linear-scaling version of the COSMO solvent model [9],[10]. COSMO is an implicit solvent model which introduces a perturbation to the semiempirical Hamiltonian that always increases the net dipole. The macrodipole calculated with the PM5 method corresponds to 492 D while the other NDDO methods give values of 478 (PM3) and 473 D (AM1). Let us compare these computed values of the macrodipole of αCT against experiments as well as against other theoretical estimates. The best set of measurements has been performed by Antosiewicz and Porschke [16]. By employing the electric dichroism technique, they found that αCT at pH 7 possesses a macrodipole of 420 D. This value is somewhat lower that those computed with our LS-SQM methods in combination with the COSMO model. A macrodipole with a magnitude of about 500 D, however, has been computed by Antosiewicz with the finite-difference Poisson-Boltzmann method [17]. This indicates that the experimental value obtained at pH 7 might be somewhat underestimated. Indeed, after re-fitting the experimental data we find that the value extrapolated at pH 7.0 corresponds to 491 D. This value is now in excellent agreement with the macrodipole computed by us with the PM5-COSMO method and displayed on the left side of Figure 4.

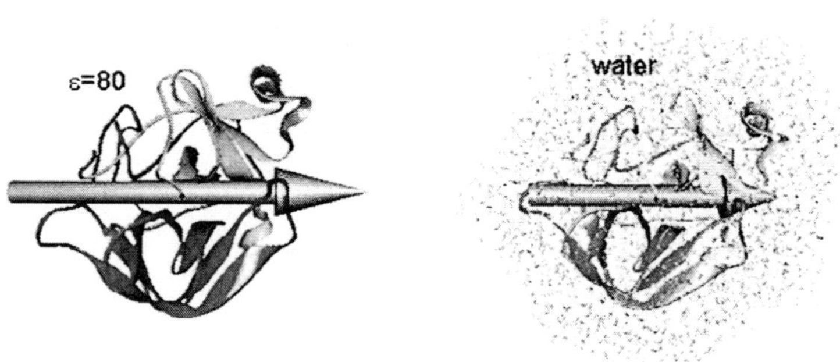

Figure 4. Macrodipole of α-chymotrypsin as computed with (left) implicit and (right) explicit solvent models

Another set of computations were performed by employing an explicit solvent model. Here a thin layer containing discrete number of water molecules is being built around the protein. Three models, each of which containing 1212, 2085, and 2811 water molecules, respectively, were built. The structure of each protein-water model was energy minimized with 1000 steps using the CFF-MM method. Subsequently, a set of QM calculations were performed on each model. The best results were obtained with the second model, which contains 2085 water molecules. The macrodipole calculated for it corresponds to 371 D (PM5 method) and is displayed on the right side of Figure 4. The remainder models are characterized by macrodipoles of lower magnitude. Several reasons can be at the origin of such results. First, the exact number of water molecules that contribute to the magnitude of the macrodipole is unknown. Second, the thickness of the water layer surrounding the real protein is likely to be not uniform since water will tend to accumulate nearby the charged residues on the protein surface. Third, protein-water interactions are not perfectly accounted for with semiempirical methods. Let us note, however, that both the orientation and direction of the dipole vectors computed with implicit and explicit solvent models agree with each other, as shown in Figure 4. This result is quite important since only the magnitude of the dipole moment vector of a protein can be experimentally determined whereas both the orientation and direction can be obtained only after performing a theoretical calculation that employs the 3D structure of the molecule.

3.2. Molecular Recognition of Sialyl Lewis X by E-selectin

Selectins are membrane proteins that are able to bind the Sialyl Lewis X (SLX) epitope, a tetrasaccharide found in glycoproteins located on leukocytes' surfaces, through their carbohydrate recognition domains (CRD). Upon skin injury, an inflammatory reaction takes place

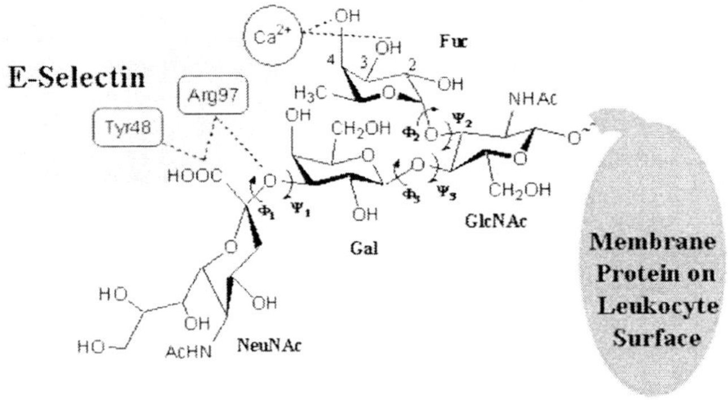

Figure 5. Chemical structure of the Sialyl Lewis X (SLX) epitope and its binding mode to E-selectin

that triggers the adhesion of leucocytes onto endothelial cells via CRD-SLX interactions. Given the importance of such biological process for medicinal chemistry, researchers have made lot of efforts to obtain structural information about this molecular recognition event. In 1994 Graves and coworkers [18] determined the structure of E-selectin by x-ray crystallography (PDB code: 1ESL). Two important amino acid residues are found on the CRD of this protein: an arginine residue (Arg97) and a tyrosine residue (Tyr48), both shown in Figure 5. Medicinal chemists have suggested earlier as the former residue, Arg97, may be involved in the formation of a salt-bridge with the carboxylate groups of SLX, as shown in Figure 6 [19]. This hypothesis, however, could not be confirmed by Graves' structure since the tetrasaccharide was not present in it. Furthermore, a single calcium ion is also always present on the CRD and deemed necessary for the recognition of the sugar. The mystery surrounding the binding mode of SLX with E-selectin was solved in 2000 when Somers et al. [20] published the crystal structure of the E-selectin: SLX complex (PDB code: 1G1T).

Although hydrogen atoms were not located in this structure, the analysis of the interatomic distances indicated that an interaction different from the one proposed earlier was likely to take place at the protein-sugar interface. As shown on the right side of Figure 6, a hydrogen bond between the NH moiety of Arg97 and the oxygen atom of the carboxyl group is expected to be formed. What remains unknown, however, is the protonation state of the carboxyl group of SLX (see Figure 5).

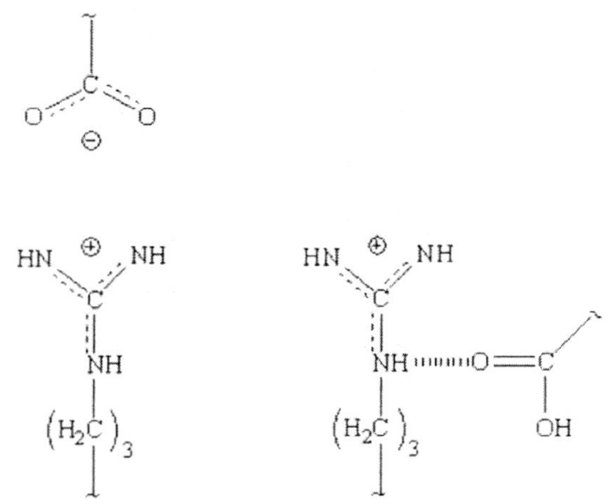

Figure 6. Interaction between the carboxyl group of SLX and Arg97: (left) salt-bridge and (right) H-bond

To shed light on this problem, we have undertaken a computational study of the SLX molecule with the aid of semiempirical MO calculations [12]. Given the size of the SLX molecule (110 atoms), we can treat it with canonical MOs. Solvation effects were introduced with COSMO by setting the dielectric constant at the value corresponding to liquid water. The starting SLX structure is that of a bioactive conformer in complex with the rat serum mannose binding protein (PDB code: 2KMB) as determined by x-ray crystallography. According to early NMR experiments on SLX in water, three low-energy conformational families, denoted to as A, B, and C, respectively, has been observed by Imberty and coworkers [21]. They differ mainly in the torsion angles about the NeuNAc(2-3)Gal glycosidic linkage. Conformer B is the one with the lowest energy while conformer A is the one that is considered to be biologically active. We will limit our discussion to the neutral and anionic forms of the latter conformer, whose structures are well reproduced with the AM1-COSMO method.

In order to compare the neutral and ionic forms of SLX, several properties were computed at the semiempirical AM1-COSMO level of theory, namely the dipole moment, electrostatic potential, conformational flexibility of the NeuNAc residue, and frontier orbitals [12]. Among them, the topology of the frontier orbitals is the most telling of all. As we can see from Figure 7, both the SLX anion and its neutral form have the HOMO localized on the NeuNAc residue. However, the LUMO is localized in different parts of the molecular skeleton. In the anion it is localized on the peptide moiety attached to GlcNAc whereas in the neutral form it is localized both on the carboxylic residue and on the oxygen atom of

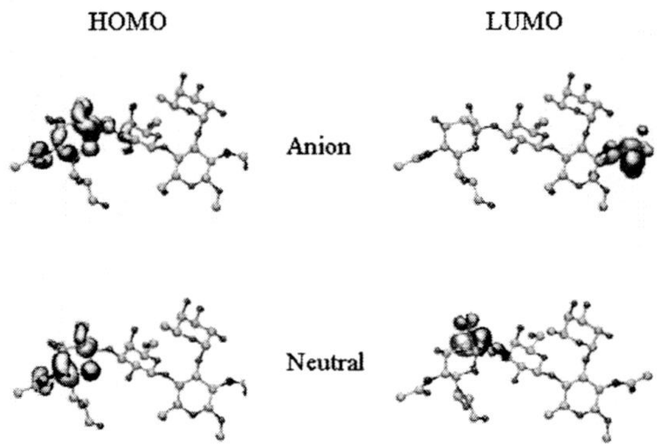

Figure 7. Frontier orbitals HOMO and LUMO of SLX in the (top) anionic and (bottom) neutral forms

the glycosidic linkage connecting NeuNAc to Gal. Interestingly, we notice that these two moieties are involved in the formation of H-bonds with the Arg97 and Tyr48 amino acid residues on the CRD. Hence, the present quantum chemical study indicates that SLX is likely to interact with the CRD of E-selectin in the neutral rather than in the anionic state. This should have important implications for the design of novel SLX analogues.

3.3. The SH2 Domain in Signal Transduction Biology

Signal transduction (ST) biology is concerned with the study of cell-cell communication, particularly the transmission of signals from the outside to the inside of the cell [22]. The interaction of a ligand, such as a hormone, with its target receptor located on the cell surface triggers a cascade of biochemical events inside the cell itself. These events do include molecular processes such as conformational changes, protein-ligand and protein-protein interactions, enzymatic reactions, and so on. Given the importance of such molecular events for ST biology, we were intrigued to study them with the aid of quantum mechanical methods. For this purpose, we chose the Src homology 2 (SH2) domain, shown in Figure 8, which is composed of about 100 amino acid residues and is found in several protein kinases that are involved in ST pathways. We selected to study the SH2 domain of human p56 lck kinase, whose structure has been determined at atomic resolution by Tong et al. [23]. This small protein contains a central antiparallel β-sheet flanked by two α-helices. A short phospho-peptide denoted to as pYEEI (with pY=phosphotyrosine, E=glutamic acid, and I=isoleucine) has been co-crystallized together with the SH2 domain. This complex represents a model of the interaction process that occurs upon the molecular recognition of phosphorylated protein chains inside the cell, which are key events of ST biology. As shown in Figure 9, the pY (or pTyr) residue makes six H-bonds with the amino acid chains located in the binding pocket of the SH2 domain. Two additional H-bonds are formed between Arg134 and acetyl capping group of pYEEI. Also, we notice the formation of five salt-bridges, three of them involving the phospho-peptide.

We have investigated the SH2:pYEEI complex and its component parts with the aid of LS-SQM calculations [13]. The AM1, PM3, and PM5 methods have been employed

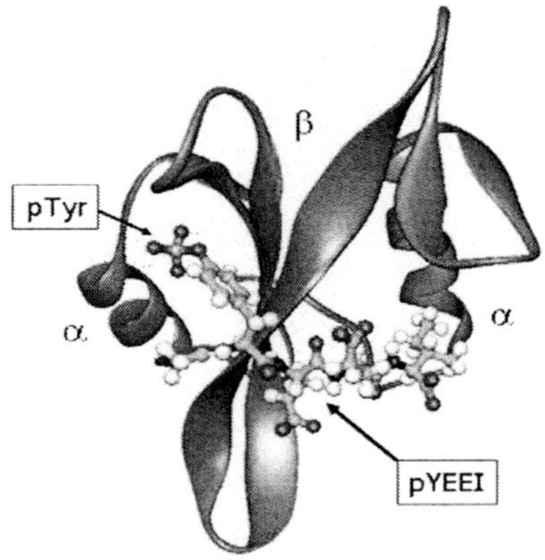

Figure 8. 3D structure of the SH2 domain in complex with the pYEEI phospho-peptide

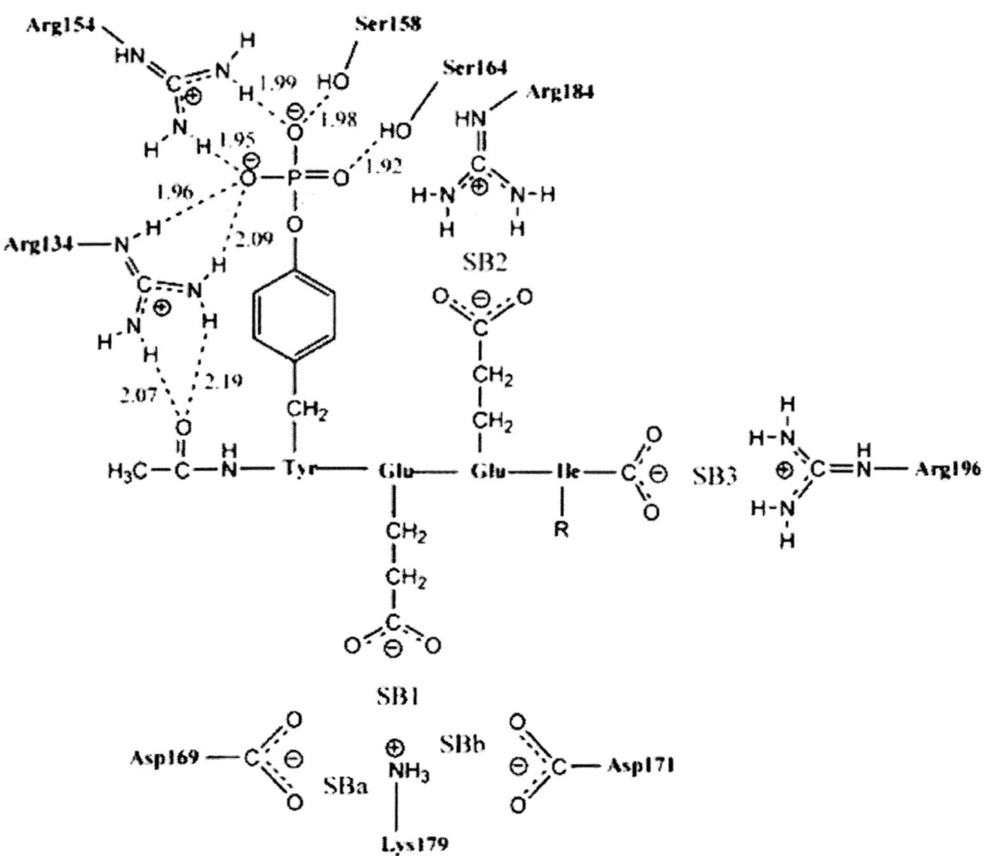

Figure 9. Intermolecular interactions of the pYEEI phospho-peptide with the amino acid residues of the SH2 domain

along with the LS version of the COSMO solvent model. Several properties have been computed: the charges of the amino acid residues located in the binding pocket of SH2, frontier orbitals, binding energy, and the dipole moment. The most interesting result concerns the change in the dipole moment of SH2 domain upon binding the phospho-peptide. As shown in Figure 10, the macrodipole of the unliganded protein corresponds to 136 D.

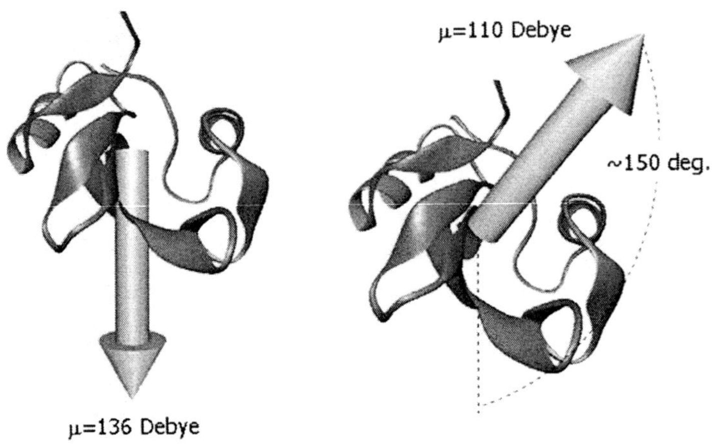

Figure 10. Ligand-induced rotation of the SH2 macrodipole: (left) unliganded and (right) liganded state

Upon binding, the dipole vector rotates by about 150 degrees while its magnitude changes only slightly to 110 D. This indicates that, upon binding, the phospho-peptide induces a change in the polarization of the protein which may have important effects in the transmission of signals inside the cell. To the best of our knowledge, this finding represents the first evidence of the occurrence of QM effects in ST biology. Furthermore, we suggest as this molecular switch could find interesting applications in the field of nanoelectronics.

4. Summary

We have discussed the results of electronic structure calculations with linear-scaling semiempirical quantum mechanical methods as applied to large biomolecules. These computational methods are complementary to classical molecular dynamics simulation and, when used together, they represent a formidable tool that allows one to shad light on complex problems that are relevant to biology and biophysics.

Acknowledgments

I thank Fujitsu Ltd. (Tokyo) for providing the MOPAC2002 software package and Dr. J.J.P. Stewart for helpful comments. The research presented here has been sponsored by the 21st century Center of Excellence (COE) program "Giant Molecules and Complex Systems" of MEXT hosted at Tohoku University.

References

[1] R. Glaser, *Biophysics*, Springer-Verlag, 2001.

[2] A. Leach, *Molecular Modelling: Principles and Applications*, Prentice Hall, 2001.

[3] W. Thiel, *Adv. Chem. Phys.*, (1996), 93, 703.

[4] J.J.P. Stewart, *Int. J. Quantum Chem.*, (1996), 58, 2744.

[5] M.J.S. Dewar, W.J. Thiel, *J. Am. Chem. Soc.*, (1977), 99, 4899.

[6] M.J.S. Dewar, E.G. Zoebish, E.F. Healy, J.J.P. Stewart, *J. Am. Chem. Soc.*, (1985), 107, 3902.

[7] J.J.P. Stewart, *J. Comput. Chem.*, (1989), 10, 209.

[8] J.J.P. Stewart, *MOPAC2002 (version 1.0)*, Fujitsu Ltd. (Tokyo).

[9] A. Klamt, G. Schürmann, *J. Chem. Soc.*, Perkin Trans. II, 1993, 799.

[10] D.M. York, T.-S. Lee, W. Yang, *Chem. Phys. Lett.*, (1996), 263, 297.

[11] F. Pichierri, *J. Mol. Struct.* (Theochem), (2003), 664-665 197.

[12] F. Pichierri, Y. Matsuo, *Bioorg. & Med. Chem.*, (2002), 10, 2751.

[13] F. Pichierri, *Biophys. Chem.*, (2004), 109, 295.

[14] D. Porschke, *Biophys. Chem.*, (1997), 66, 241.

[15] H. Tsukada, D.M. Blow, *J. Mol. Biol.*, (1985), 184, 703.

[16] J. Antosiewicz, D. Porshke, *Biochemistry*, (1989), 28, 10072.

[17] J. Antosiewicz, *Biophys. J.*, (1995), 69, 1344.

[18] B.J. Graves et al., *Nature*, (1994), 367, 4431.

[19] E.E. Simanek et al., *Chem Rev.*, (1998), 98, 833.

[20] W.S. Somers et al., *Cell*, (2000), 103, 467.

[21] A. Imberty, S. Peres, *Chem. Rev.*, (2000), 100, 4567.

[22] K.M. Eyster, *Biochem. Pharmacol.*, (1998), 55, 1921.

[23] L. Tong et al., *J. Mol. Biol.*, (1996), 256, 601.

In: Molecular Simulation Studies in Material ...
Editor: Kholmirzo Kholmurodov, pp. 15-29
ISBN 1-59454-607-X

Chapter 3

COMPUTER SIMULATIONS OF MODEL LIPID MEMBRANES

A. L. Rabinovich and P. O. Ripatti*
Institute of Biology, Karelian Research Centre, Russian Academy of Sciences,
11 Pushkinskaya str., Petrozavodsk 185910, Russia
N. K. Balabaev†
Institute of Mathematical Problems of Biology, Russian Academy of Sciences,
Pushchino, Moscow Region, 142290, Russia

Abstract

Molecular dynamics (MD) simulations using the Collision Dynamics method are described, for several model membranes:

i. hydrated homogeneous lipid bilayers constructed of saturated and unsaturated phosphatidylcholines (PC), – 1-stearoyl-2-oleoyl-*sn*-glycero-3-PC, 1-stearoyl-2-linoleoyl-*sn*-glycero-3-PC, 1-stearoyl-2-linolenoyl-*sn*-glycero-3-PC, 1-stearoyl-2-arachidonoyl-*sn*-glycero-3-PC, 1-stearoyl-2-docosahexaenoyl-*sn*-glycero-3-PC;

ii. the above-listed PC bilayers with a cylinder hole and the PC bilayers with a peptide inclusion;

iii. the PC bilayers with several sphere-cylindrical holes and the PC bilayers with a low mole fraction of cholesterol.

Several results of the investigations are discussed for the homogeneous bilayers. The $C-H$ and $C-C$ bond order parameter $-S_{CH}$, $-S_{CC}$ profiles of the hydrocarbon tails, the bond orientation distribution functions and the root-mean-square values of the positional fluctuations of the lipid chain carbons, mass density profiles along the bilayer normal (Z) for different atoms and groups of atoms were calculated. In the homogeneous bilayers the order parameter characteristics of lipid hydrocarbon chains in the "liquid" region of the bilayer (somewhat remote from the membrane-water interface) are found to be qualitatively similar to those of single unperturbed hydrocarbon chains:

*E-mail address: [rabinov,ripatti]@krc.karelia.ru
†E-mail address: balabaev@impb.psn.ru

the behavior of the acyl chains in this region is dominated by the intramolecular short-range interactions. A close relationship between the physical characteristics and the structure of the chains is elucidated: the number of the chain carbons, the number of cis double bonds and their position in the chain determine principally the calculated properties.

1. Introduction

Biological membranes are enormously complex objects both in their structure and dynamical properties; they consist of molecules of lipids, proteins, carbohydrates. A typical biomembrane contains many species of lipid molecules, with different head groups and hydrocarbon chains. At least half of the fatty acid (FA) chains in most cases are unsaturated. The double bonds of natural polyunsaturated (PU) FA chains are, as a rule, separated by a methylene group; i.e., the double bonds are methylene-interrupted. Unsaturated and polyunsaturated lipids are of significant functional importance in biomembranes. There is an extensive literature concerned with FA chain composition in relation to the roles of the cis double bonds in membranes (see, e.g., [1, 2, 3, 4]). An important component of biomembranes is cholesterol [5, 6, 7]. The presence of cholesterol is required for normal functioning of mammalian plasma membranes; cholesterol modulates phospholipid organization within membranes, but little is known about the interaction of cholesterol with polyunsaturated phospholipids. Another important topics of the investigations are interactions between lipids and different proteins or peptides [8, 9, 10, 11], – in particular, Gramicidin A, a linear hydrophobic membrane pentadecapeptide with channel-forming properties [12]. The effects of Gramicidin A on the structure of phospholipid assemblies are of great interest. The detailed study of the properties of phospholipid bilayer membranes with different inclusions is an important step toward the understanding of the molecular basis of many biological processes.

Computer simulation methods, such as Monte Carlo (MC) or molecular dynamics (MD) [13, 14, 15] have been widely used for studying equilibrium and dynamical properties of various complex biological systems, including lipid bilayers [16]-[26]. In the recent years several unsaturated and PU lipid systems have been studied by computer simulations (for reviews, see [27, 28, 29, 30, 31]).

In this paper, MD simulations using the Collision Dynamics method are described, of several model membranes:

i. hydrated homogeneous lipid bilayers constructed of saturated and unsaturated phosphatidylcholines (PC), –
   1-stearoyl-2-oleoyl-*sn*-glycero-3-PC (18:0/18:1 PC),
   1-stearoyl-2-linoleoyl-*sn*-glycero-3-PC (18:0/18:2 PC),
   1-stearoyl-2-linolenoyl-*sn*-glycero-3-PC (18:0/18:3 PC),
   1-stearoyl-2-arachidonoyl-*sn*-glycero-3-PC (18:0/20:4 PC),
   1-stearoyl-2-docosahexaenoyl-*sn*-glycero-3-PC (18:0/22:6 PC);

ii. the above-listed PC bilayers with a cylinder hole and the PC bilayers with a peptide inclusion;

iii. the PC bilayers with several sphere-cylindrical holes and the PC bilayers with a low mole fraction of cholesterol.

Several results of the investigations of homogeneous lipid bilayers are discussed.

The above-mentioned lipid acyl chains are typical for natural membranes. We will use the notation $N : k$ for the FA chain structure, where N refers to the total number of carbon atoms in the chain, k is the number of (methylene-interrupted) double bonds.

2. Homogeneous PC Bilayers: MD Simulations (Collision Dynamics Method)

The main idea of MD simulations of a many-particle system is the solution of the Newton's equations of motion for a set of particles (atoms or molecules) that comprises the system. This procedure includes the description of the atomic system model, atom-atom interaction potentials, boundary conditions of the system and an approximate step-by-step technique for solving the classical equations of motion.

The simulation boxes of the bilayers have contained 96 PC molecules and 2304 water molecules: 48 lipid molecules per layer and 24 H_2O molecules per lipid in each case [27, 28, 29, 32]. Each lipid molecule has contained a saturated (18:0) chain and an unsaturated one. The water phase has consisted of explicit TIP3P water [33] molecules. The H_2O geometry [33] parameters, the partial charges [33] and the force constants [34] for H_2O have been used.

The potential energy U_{bil} of a homogeneous lipid bilayer has been calculated as

$$U_{bil} = U_b + U_a + U_t + U_{oop} + U_{vdw} + U_e, \tag{1}$$

where U_b is the bond-stretching energy:

$$U_b = \sum K_l (l - l_0)^2, \tag{2}$$

U_a is the angle-bending energy:

$$U_a = \sum K_\Theta (\Theta - \Theta_0)^2, \tag{3}$$

U_t is the torsion energy:

$$U_t = \sum K_\varphi [1 + \delta \cos(n_0 \varphi)], \tag{4}$$

U_{oop} is the out-of-plane energy (only for the double bonds and the carbonyl groups):

$$U_{oop} = \sum K_\phi [1 - \cos(2\phi)], \tag{5}$$

U_{vdw} is the van der Waals energy: $U_{vdw} = \sum\sum U_{LJ}(r_{ij}) W_{vdw}(r_{ij})$, where

$$U_{LJ} = 4\varepsilon_{ij}[(\sigma_{ij}/r_{ij})^{12} - (\sigma_{ij}/r_{ij})^6] \tag{6}$$

and W_{vdw} is the switching function:

$$W_{vdw}(r_{ij}) = \begin{cases} 1, & r_{ij} \le R_{on}; \\ \dfrac{(R_{off}^2 - r_{ij}^2)^2 (R_{off}^2 - 3R_{on}^2 + 2r_{ij}^2)}{(R_{off}^2 - R_{on}^2)^3}, & R_{on} < r_{ij} < R_{off}; \\ 0, & r_{ij} \ge R_{off}. \end{cases} \tag{7}$$

U_e is the electrostatic energy:

$$U_e = \sum\sum [q_i q_j/(\varepsilon r_{ij})] W_e(r_{ij}) \tag{8}$$

where W_e is the screening function:

$$W_e(r_{ij}) = \begin{cases} (1 - r_{ij}/R_e)^2, & r_{ij} \le R_e; \\ 0, & r_{ij} > R_e. \end{cases} \tag{9}$$

In eqns 1 to 9 the following notations are used: l is the bond length, Θ is the bond angle, φ is the torsion angle, ϕ is the out-of-plane angle, l_0, Θ_0 are equilibrium values for the bond lengths and angles; K_l, K_Θ, K_φ, K_ϕ are force constants for the bonds, angles, dihedrals and out-of-plane angles, respectively; n_0 is the dihedral multiplicity; $\delta = \pm 1$; r_{ij} is the distance between nonbonded atoms i and j; ε_{ij}, σ_{ij} are the Lennard-Jones parameters for the atom pairs, R_{on}, R_{off} are the switching parameters; q_i, q_j are the partial charges on atoms i, j; ε is the dielectric constant; R_e is the screening radius.

Structural parameters for lipid molecules (bond lengths, angles, force constants) and torsion parameters have been generally chosen according to the set [34] (which is close to the parameter set of the work [35]) and the structural data of the works [36, 37] have been also taken into account. The partial charges [35] (in electron units) have been used with consideration for the data of unsaturated lipid chains [38]. The parameters for nonbonded (van der Waals) interactions have been borrowed from [39].

The equations of motion have been integrated by using the "velocity Verlet" algorithm [13, 15]. The simulation time step has been 10^{-15} sec. The Collisional Dynamics [40] (CD) method has been used, i.e., the equations of motion (for the NPT-ensemble) have been integrated as follows:

$$\frac{dr_{i,\alpha}}{dt} = v_{i,\alpha} + \chi_\alpha \beta_p (P_\alpha - P_{\alpha,ref}) r_{i,\alpha}, \tag{10}$$

$$m_i \frac{dv_{i,\alpha}}{dt} = -\frac{\partial U_{bil}}{\partial r_{i,\alpha}} + \sum_k f_{k,\alpha}^i \delta(t - t_k^i), \tag{11}$$

where $i = 1, 2, \ldots, N$, $\alpha = \{x, y, z\}$, $\delta(t)$ is the Dirac delta function, $f_{k,\alpha}^i$ and t_k^i $(k = 1, \ldots)$ are the impulse forces and accidental time moments of atom collisions with the virtual bath particles, $r_{i,\alpha}$, $v_{i,\alpha}$ are the coordinates and velocities of the i-th atom, β_p, χ_α are parameters for the pressure relaxation, $P_{\alpha,ref}$ are the reference values of the pressure P_α components.

The CD thermostat has been implemented by stochastic collisions with virtual particles of mass m_0 :

$$\vec{f}_k = 2\tfrac{m_0 m_i}{m_0 + m_i} (\vec{v}_0 - \vec{v}_i),$$

$v_{0,\alpha}$ are chosen from a Gaussian distribution

$$P(\vec{v}_0) = \left(\frac{m_0}{2\pi k_B T_{ref}}\right)^{3/2} \cdot \exp\left(-\frac{m_0 \vec{v}_o^2}{2k_B T_{ref}}\right), \quad (12)$$

where T_{ref} is the reference temperature and k_B is the Boltzmann constant. The collisions occur in accordance with a Poisson process specified by a single parameter λ_0 which is the mean number of collisions that atom i suffers per unit time ("collision frequency"). All the details of the CD-method are described previously [40].

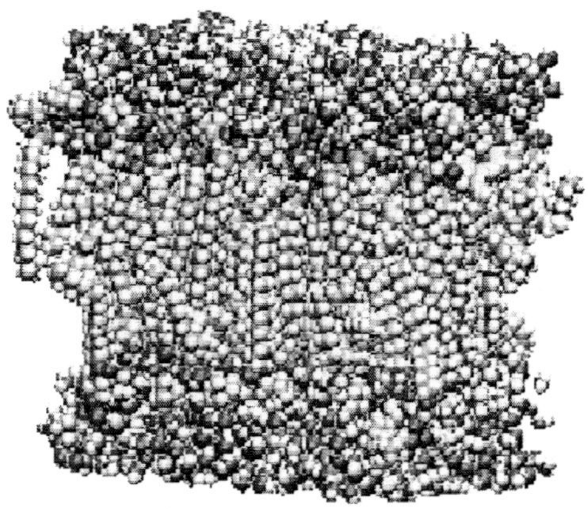

Figure 1. A typical snapshot of the hydrated 1-stearoyl-2-arahidonoyl-*sn*-glycero-3-phosphatidylcholine (18:0/20:4 PC) bilayer - along the Molecular Dynamics trajectory; 140 atoms per lipid, 96 lipids and 2304 water molecules in the box with periodic boundary conditions (altogether there are 20352 atoms). Color coding is as follows. Carbon atoms are green, oxygens are red, double bonded carbons are yellow, hydrogens are grey, phosphorus atoms are black, nitrogens are dark-blue

Periodic boundary conditions in the three directions (x, y, z) have been applied. The bilayer systems have been coupled to an external temperature bath T_{ref} = 303 K and to the reference pressure $P_{\alpha,ref}$ = 1 atm. The simulation temperature of 303 K is higher than the relevant experimental gel-to-liquid crystal phase transition temperature T_c [41] for these PCs. A typical snapshot of a bilayer configuration is presented in Fig.1. The trajectories of $\sim$1 ns have been calculated for the 5 bilayers.

3. Circular Cylindrical Hole Creation in Homogeneous PC Bilayers

The elementary model of an inclusion in a bilayer is a circular cylindrical hole (cavity), the axis of the cylinder (with the coordinates X_{axis}, Y_{axis}) is parallel to the bilayer normal Z. To create the box of a PC bilayer with a cylindrical hole in the course of the MD simulation the calculation box of the homogeneous PC bilayer has been used. The technique is as follows.

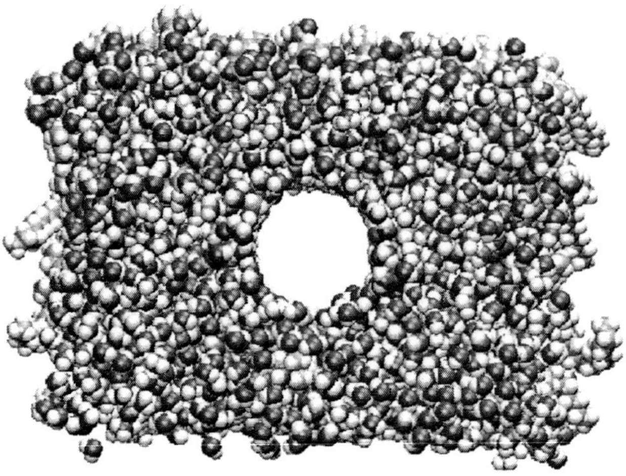

Figure 2. A typical snapshot for a lipid bilayer with a cylindrical hole along the MD trajectory (top view). Color coding – as in Fig.1

The interaction energy $U_{i,R}$ of the i-th atom of the lipid or water molecules and the cylinder has been calculated as

$$U_{i,R}(r_{i,xy}) = \begin{cases} \varepsilon_{cyl} \cdot \left[\left(\frac{R_{minh}}{r_{i,xy}-R} \right)^{12} - 2 \cdot \left(\frac{R_{minh}}{r_{i,xy}-R} \right)^{6} \right] + \varepsilon_{cyl}, & |r_{i,xy} - R| < R_{minh}; \\ 0, & |r_{i,xy} - R| \geq R_{minh}. \end{cases} \quad (13)$$

Here

$$r_{i,xy} = \left[(x_i - X_{axis})^2 + (y_i - Y_{axis})^2 \right]^{1/2},$$

x_i, y_i are coordinates of atom i, R is the radius of the cylinder, ε_{cyl} and R_{minh} are parameters.

To create the hole in the bilayer the radius R of the cylinder has been given as

$$\begin{cases} R = -R_{minh} + V_R t, & t \leq t_R; \\ R = R_{fix}, & t > t_R, \end{cases}$$

where $V_R = Const$ is the "velocity" of R, t is the time, R_{fix} is the prescribed value of R,

$$t_R = (R_{fix} + R_{minh}) / V_R.$$

The potential energy U_{cyl} of a bilayer with the cylindrical hole instead of the energy U_{bil} has been used for the MD simulation according to eqns 10 and 11, where

$$U_{cyl} = U_{bil} + \sum U_{i,R},$$

the energy U_{bil} is from eqn 1, $\sum U_{i,R}$ is the energy of the interaction of all the atoms i (of the lipids and water) with the cylindrical hole. Then each bilayer system has been brought to equilibrium and the trajectory of $\sim$1 ns has been calculated (Fig.2).

Real inclusions with the explicit structure (e.g., a protein or peptide molecule) can be then inserted into the hole of a given bilayer. The best structurally and functionally characterized "molecular pore" is currently Gramicidin A: this pentadecapeptide adopts a head-to-head β – helical dimer conformation to form an ion-conducting narrow pore [12]. Fig.3 demonstrates a dimer of Gramicidin A inserted into a PC bilayer.

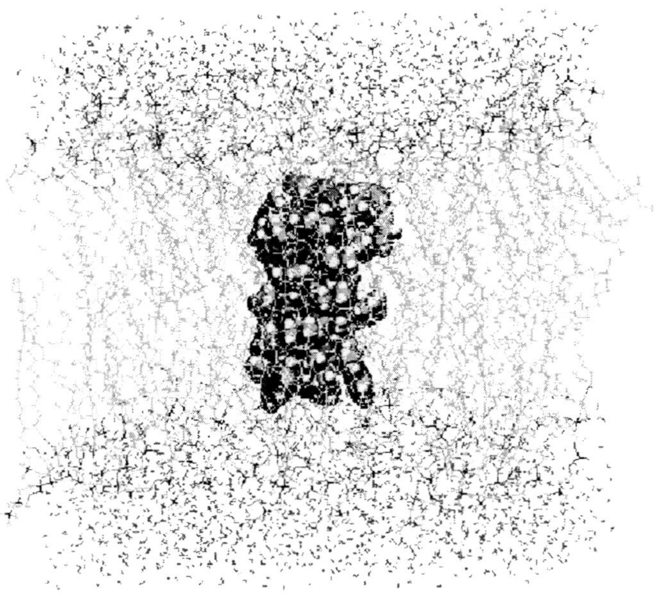

Figure 3. A typical snapshot of the hydrated 1-stearoyl-2-docosahexaenoyl-*sn*-glycero-3-PC (18:0/22:6 PC) bilayer with a dimer of Gramicidin A along the MD trajectory: 21996 atoms in the box with periodic boundary conditions. The lipid and water molecules are represented by the wire model, Gramicidin A – by the spacefilling model, the color coding – as in Fig.1

4. Sphere-Cylindrical Holes Creation in Homogeneous PC Bilayers

As in the case of a circular cylindrical hole the calculation box of the homogeneous PC bilayer has been used to create the box of a PC bilayer with several sphere-cylindrical holes in the course of the MD simulation. The procedure is as follows. The axes of the sphere cylinders with the coordinates $X^{(k)}_{axis}$, $Y^{(k)}_{axis}$ are parallel to the bilayer normal Z, where $k = 1, 2, \ldots, p$; p is the total number of the sphere cylinders; $Z^{(k)}_0$ are coordinates of the centres of the sphere cylinders along their axes. The interaction energy $U^{(k)}_{i,hole}$ of the i-th atom of the lipid or water molecules and the k-th sphere-cylindrical hole has been calculated as

$$U^{(k)}_{i,hole} = \begin{cases} U^{(k)}_{i,R}\left(r^{(k)}_{i,xy}\right), & Z^{(k)}_{-} \le z_i \le Z^{(k)}_{+}; \\ U^{(k)}_{i,R}\left(r^{(k)}_{i,xyz}\right), & Z^{(k)}_{+} < z_i < Z^{(k)}_{-}, \end{cases} \tag{14}$$

where

$$U^{(k)}_{i,R}(r) = \begin{cases} \varepsilon^{(k)}_{hole} \cdot \left[\left(\frac{R^{(k)}_{minh}}{r-R^{(k)}}\right)^{12} - 2\cdot\left(\frac{R^{(k)}_{minh}}{r-R^{(k)}}\right)^{6}\right] + \varepsilon^{(k)}_{hole}, & |r-R^{(k)}| < R^{(k)}_{minh}; \\ 0, & |r-R^{(k)}| \ge R^{(k)}_{minh}. \end{cases} \tag{15}$$

The argument r of the function $U_{i,R}^{(k)}(r)$ is equal to $r_{i,xy}^{(k)}$ for the cylinder region of the k-th sphere cylinder and to $r_{i,xyz}^{(k)}$ – for the regions of the two hemispheres.

Here

$$r_{i,xy}^{(k)} = \left[\left(x_i - X_{axis}^{(k)}\right)^2 + \left(y_i - Y_{axis}^{(k)}\right)^2\right]^{1/2}$$

is the distance between the i-th atom and the axis of the k-th sphere cylinder;

$$r_{i,xyz}^{(k)} = \left[\left(x_i - X_{axis}^{(k)}\right)^2 + \left(y_i - Y_{axis}^{(k)}\right)^2 + \left(z_i - Z_{\pm}^{(k)}\right)^2\right]^{1/2}$$

is the distance between the i-th atom and the centre of the first or second hemisphere of the k-th sphere cylinder; x_i, y_i, z_i are coordinates of the i-th atom of the lipid or water molecules; $R^{(k)}$ is the radius of the k-th sphere cylinder; $\varepsilon_{hole}^{(k)}$ and $R_{minh}^{(k)}$ are parameters; $Z_{-}^{(k)}$, $Z_{+}^{(k)}$ are coordinates of the minimal and maximal points at the height of cylindrical part of the k-th sphere cylinder, i.e.,

$$lZ_{-}^{(k)} = Z_0^{(k)} - H^{(k)}/2,$$
$$Z_{+}^{(k)} = Z_0^{(k)} + H^{(k)}/2,$$

where $H^{(k)}$ is the height of the k-th sphere cylinder.

To create all the holes in the bilayer the following time dependencies of the radius $R^{(k)}$ and height $H^{(k)}$ of the k-th sphere cylinder have been used:

$$\begin{cases} R^{(k)} = -R_{minh}^{(k)} + V_R^{(k)} t, & t \le t_R^{(k)}; \\ R^{(k)} = R_{fix}^{(k)}, & t > t_R^{(k)}; \end{cases}$$

$V_R^{(k)} = Const$ is the "velocity" of the radius $R^{(k)}$ changes, $R_{fix}^{(k)}$ is the prescribed value of the radius;

$$t_R^{(k)} = (R_{fix}^{(k)} + R_{minh}^{(k)})/V_R^{(k)};$$

$$\begin{cases} H^{(k)} = H_0^{(k)} + V_H^{(k)} t, & t \le t_H^{(k)}; \\ H^{(k)} = H_{fix}^{(k)}, & t > t_H^{(k)}, \end{cases}$$

where $H_0^{(k)}$ is the initial height, $V_H^{(k)} = Const$ is the "velocity", and $H_{fix}^{(k)}$ is the prescribed value of $H^{(k)}$,

$$t_H^{(k)} = (H_{fix}^{(k)} - H_0^{(k)})/V_H^{(k)}.$$

The potential energy U_{hole} of all the lipid and water atoms and all the k sphere-cylindrical holes has been equal to

$$U_{hole} = \sum_k \sum_i U_{i,hole}^{(k)}.$$

The equations of motion 10 and 11 have been used for the MD simulation, with the energy

$$U_{spcyl} = U_{bil} + U_{hole}$$

instead of U_{bil}. To investigate the mixed lipid bilayers (i.e., the bilayers with any real small "guest" molecules) each of the obtained bilayer systems can be used as a matrix, any real molecular inclusion can be inserted into the k-th hole.

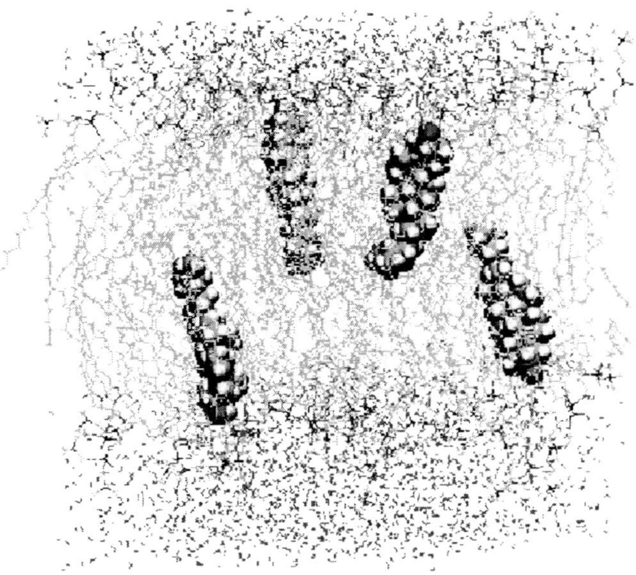

Figure 4. A typical snapshot of the hydrated 1-stearoyl-2-arahidonoyl-sn-glycero-3-PC (18:0/20:4 PC) bilayer with cholesterol along the MD trajectory: 96 lipid, 4 cholesterol and 2400 water molecules in the box with periodic boundary conditions (20936 atoms). The lipid and water molecules are represented by the wire model, cholesterol – by the spacefilling model. Color coding – as in Fig.1

Fig.4 illustrates an example of the obtained PC bilayer with a low mole fraction of cholesterol.

5. Results and Discussion

Several physical characteristics of the bilayer systems have been calculated. Some of the results obtained only for the pure lipid bilayers are discussed below. Mass density profiles along the PC bilayer normals for different groups of atoms are shown in Fig.5 (left). In the calculations the position of the bilayer centre ($z = 0$) for all the PC bilayers has been calculated as the z-position being at equal distances from the average positions of the centres of the $P-N$ vectors of the two opposite monolayers. The average distance between the $P-N$ vectors along the normal is found to be 4.39 nm for the 18:0/18:1 PC bilayer, 4.27 nm for the 18:0/18:2 PC, 4.28 nm for the 18:0/18:3 PC, 4.26 nm for the 18:0/20:4 PC and 4.28 nm for the 18:0/22:6 PC bilayer. The average phosphorus - phosphorus distances d_{PP} along the normal are found to be 4.27, 4.18, 4.19, 4.14 and 4.18 nm, respectively. The hydrocarbon region widths (i.e., the average $C...C$ atom distances of $C=O$ groups of the acyls between the two sides of the bilayers) are found to be 3.30, 3.22, 3.25, 3.18 and 3.23 nm, respectively. The average cross-sectional areas per PC molecule during the simulations are found to be 0.6134 nm^2, 0.6242 nm^2, 0.6170 nm^2, 0.6525 nm^2 and 0.6624 nm^2, respectively. The obtained results are qualitatively in line with experimental data and the data of another computer simulations of various unsaturated lipid bilayers. Indeed, according to X-ray diffraction studies d_{PP} = 4.04 nm for the 18:1/18:1 PC bilayer [42].

According to MD simulations $d_{PP} = 4.16$ nm [43] for the 16:0/18:2 PC bilayer at $T = 321$ K (the average area per molecule is 0.7056 nm^2) and $d_{PP} = 4.25$ nm [38] for the 18:1/18:1 PC bilayer at $T = 296$ K (the average area per molecule is 0.593 nm^2).

It has been observed for the 18:0/18:1 PC, 18:0/18:2 PC, 18:0/18:3 PC and 18:0/20:4 PC bilayers, that the z-positions of the end carbons of saturated sn-1 chains (i.e., carbons no.18) of the opposite monolayers overlap. The z-positions of the end carbons of sn-2 (i.e., unsaturated) chains of the 1st and 2nd monolayers in the four bilayers are separated. It should be emphasized that the z-positions of the end carbons of sn-2 chains in the opposite monolayers of the 18:0/22:6 PC bilayer are found to be separated as well. Thus, acyl unsaturation has a strong effect on the level of chain interdigitation. The use of unsaturated lipids in the bilayer may help to regulate this property. The more lipid chains cross the mid-plane of the bilayer the stronger is the coupling between the two monolayers.

The $C-H$ and $C-C$ bond order parameter $-S_{CH}$, S_{CC} profiles given by

$$S = (3 < \cos^2 \beta > -1)/2, \tag{16}$$

have been calculated for the 5 bilayers, where β is the angle between the bond $C-H$ (or $C-C$) and the bilayer normal. With increasing lipid chain unsaturation the average magnitude $|S_{CH}|$ of the $C-H$ bond order parameter of unsaturated chain is found to decrease (Fig.5, right), similar to the unsaturated monolayers [44, 45, 46, 47, 48, 49, 50] and bilayers [27, 28, 51, 52] investigated before. Another characteristic feature in the properties of unsaturated lipids is that the order parameters S_{CC} of single $C-C$-bonds next to the cis double bond $C=C$ in the unsaturated chain are lower than that for the double bond $C=C$ and for the corresponding single bond $C-C$ in the saturated acyl chains.

Bond-vector orientation probability distribution functions (with respect to the bilayer normal) for all the $C-H$ and $C-C$ bonds of the chains of the PC bilayers and FWHM (full width at half maximum) of each distribution have been calculated. The FWHM values describe the "fluctuation" contributions to the order parameter values. One of the most important common results concerning bond orientation probability distribution functions of the PC bilayers seems to be the "broadening effect": FWHM value corresponding to the distribution function of the $C-H$ bonds of methylene group located between the cis double bonds in the PU chains is higher than that for the $C-H$ bonds of the double bonded carbons. The "broadening effect" was first detected for isolated PU hydrocarbon chains in the MC simulations [53, 54, 55, 56, 57, 58, 59], then for monolayers consisted of PU molecules of diacylglicerolipids (DGs) [44, 45, 46, 48, 49, 50] and for DG- and PC bilayers [27, 28, 60] in the MD simulations.

The root-mean-square (RMS) values of the fluctuations of the chain carbons of all the PC bilayers have been calculated: in "plane" (xy), out-of-plane (in the z-direction) and the total. The RMS fluctuations of the unsaturated chain carbons are found to be higher than those of saturated one. With increasing chain unsaturation this difference is found to increase. The RMS fluctuations of the double bonded carbons proved to be increased as compared to the adjacent single bonded carbons of the same chain (the "$C=C$-effect" [61]). The "$C=C$-effect" is in line with the "broadening effect".

The comparison of the results of the MD and MC simulations with the available experimental data has demonstrated that the order parameter characteristics of the hydrocarbon chains in the "liquid" region of the lipid monolayers and bilayers (i.e., somewhat remote

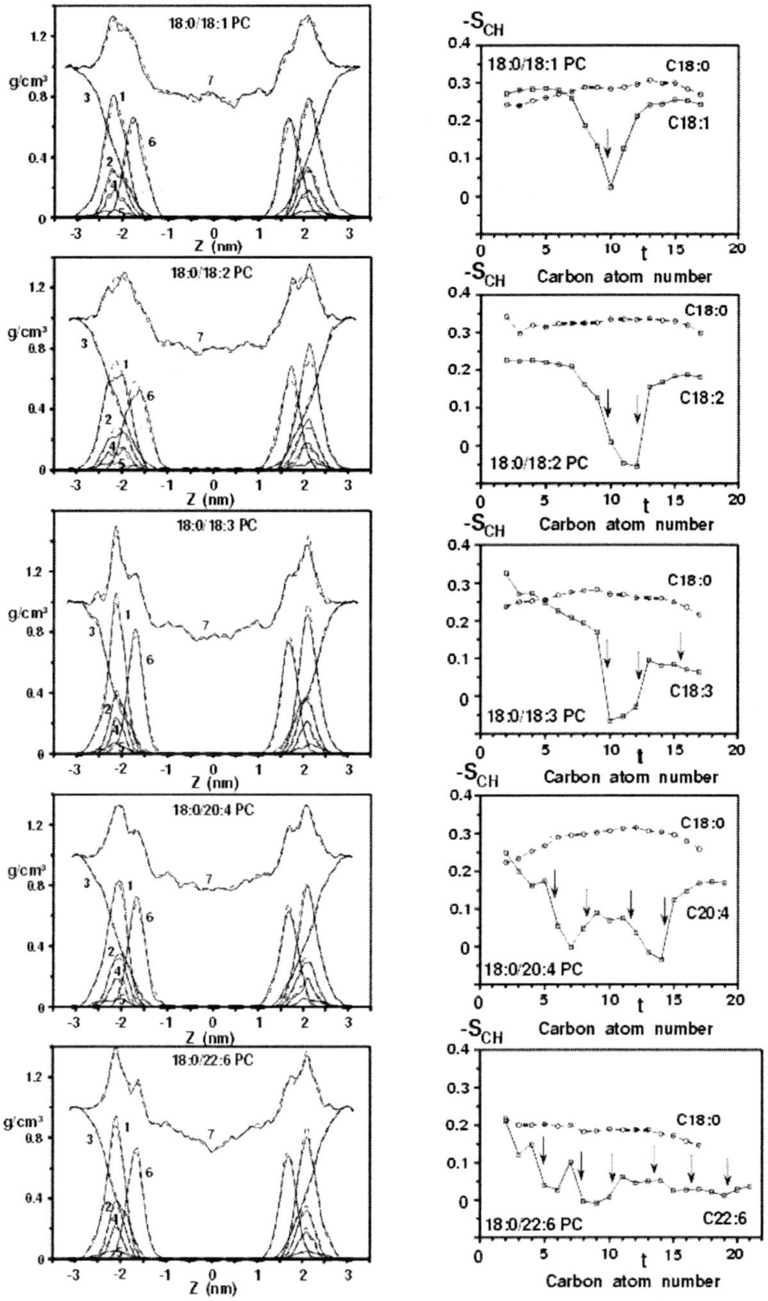

Figure 5. Left column: Mass density distributions along the normal for selected groups of atoms in the hydrated bilayers of 18:0/18:1 PC, 18:0/18:2 PC, 18:0/18:3 PC, 18:0/20:4 PC and 18:0/22:6 PC as determined from the MD simulations: phosphatidylcholine head groups (1), oxygen atoms of phosphate PO_4 groups (2), water (3), phosphorus atoms (4), nitrogen atoms (5), atoms of glycerol and groups $O-C=O$ (6), total (7). The zero point ($z=0$) is the centre of the bilayer. The dashed lines correspond to the data symmetrized over the both halves of the bilayer. Right column: Order-parameter profiles $-S_{CH}$ for both tails of the lipid molecules of the same bilayers; the arrows indicate the double bond positions

from the membrane-water interface) are qualitatively similar to those of single unperturbed [36] hydrocarbon chains. This indicates that the behavior of the lipid hydrocarbon chains in this region is dominated by the intramolecular short-range [36] interactions, because the properties of the unperturbed hydrocarbon chain are fully defined by the short-range interaction energy [36]. Thus, the long-range [36] interactions of the lipid segments in the "liquid" region of the bilayer and the interactions with atoms of the bilayer-water interface may be considered as a disturbance: the intermolecular interactions are largely used to orient the lipid molecules in the direction of the membrane normal.

The analysis of various unsaturated pure lipid bilayers has revealed that although many details are observed in the physical characteristics (such as bond ordering or fluctuations) of the atoms and bonds in the tails depending on specific conditions, the main qualitatively features of the properties (e.g., the shape of the bond order parameter profiles or RMS deviations profiles) for a given acyl chain remain unchanged, - they are like "fingerprints" pointing one-to-one to the details of the unsaturation along the chain. The number of the chain carbons, the number of cis double bonds and their position in the chain determine principally (qualitatively) the calculated properties.

Acknowledgments

This work has been supported by RFBR (grant 03-03-32229) and INTAS (grant 01-0067)

References

[1] E.A. Dratz and A.J. Deese, *In: "The health effects of polyunsaturated fatty acids in seafoods"*, A.P. Simopoulous, R.R. Kifer, R.E. Martin eds., Acad. Press, New York, 1986, 319-351.

[2] P.J. Quinn, F. Joo, and L. Vigh, *Progress in Biophysics & Molecular Biology*, (1989), 53, 71-103.

[3] A.L. Rabinovich and P.O. Ripatti, *Uspekhi sovremennoi biologii*, (1994), 114, 581-594 (*in Russian*).

[4] S. Everts and J.H. Davis, *Biophysical Journal*, (2000), 79, 885-897.

[5] H. Ohvo-Rekilä, B. Ramstedt, P. Leppimäki, and J.P. Slotte, *Progress in Lipid Research*, (2002), 41, 66-97.

[6] T.P.W. McMullen, R.N.A.H. Lewis, and R.N. McElhaney, *Current Opinion in Colloid and Interface Science*, (2004), 8, 459-468.

[7] S.A. Pandit, D. Bostick, and M.L. Berkowitz, *Biophysical Journal*, (2004), 86, 1345-1356.

[8] A.G. Lee, *Biochimica et Biophysica Acta*, (2003), 1612, 1-40.

[9] J.A. Hamilton, *Progress in Lipid Research*, (2004), 43, 177-199.

[10] M.O. Jensen, O.G. Mouritsen, and G.H. Petersz, *Biophysical Journal*, (2004), 86, 3556-3575.

[11] T.J. McIntosh, *Chemistry and Physics of Lipids*, (2004), 130, 83-98.

[12] B. Roux, *Accounts of Chemical Research*, (2002), 35, 366-375.

[13] M.P. Allen and D.J. Tildesley, *Computer Simulation of Liquids*, Clarendon Press, Oxford, 1987.

[14] D. Frenkel and B. Smit, *Understanding Molecular Simulation. From Algorithms to Applications*, Academic Press, San Diego, 1996.

[15] R.J. Sadus, *Molecular Simulations of Fluids. Theory, Algorithms and Object-Orientation*, Elsevier, Amsterdam etc., 1999.

[16] M. Karplus, *Accounts of Chemical Research*, (2002), 35, 321-323.

[17] H.J.C. Berendsen, *Science*, (2001), 294, 2304-2305.

[18] R.W. Pastor, *Current Opinion in Structural Biology*, (1994), 4, 486-492.

[19] K.V. Damodaran and K.M. Merz, *In: "Reviews in Computational Chemistry"*, K.B. Lipkowitz and D.B. Boyd eds., VCH Publishers, Inc. N.Y., 5, 269-298, 1994.

[20] D.J. Tobias, K. Tu, and M.L. Klein, *Current Opinion in Colloid & Interface Science*, (1997), 2, 15-26.

[21] D.P. Tieleman, S.J. Marrink, and H.J.C. Berendsen, *Biochimica et Biophysica Acta*, (1997), 1331, 235-270.

[22] K.M. Merz, *Current Opinion in Structural Biology*, (1997), 7, 511-517.

[23] L.R. Forrest and M.S.P. Sanson, *Current Opinion in Structural Biology*, (2000), 10, 174-181.

[24] S.E. Feller, *Current Opinion in Colloid & Interface Science*, (2000), 5, 217-223.

[25] D.J. Tobias, *In: "Computational Biochemistry & Biophysics"*, O.H. Becker, A.D. MacKerell, B. Roux, and M. Watanabe eds., Dekker, N.Y., 2000.

[26] L. Saiz and M.L. Klein, *Accounts of Chemical Research*, (2002), 35, 482-489.

[27] A.L. Rabinovich, P.O. Ripatti, N.K. Balabaev, and F.A.M. Leermakers, *Proceedings of SPIE*, (2002), 4627, 141-153.

[28] A.L. Rabinovich, P.O. Ripatti, N.K. Balabaev, and F.A.M. Leermakers, *Physical Review E*, (2003), 67, 011909.

[29] F.A.M. Leermakers, A.L. Rabinovich, and N.K. Balabaev, *Physical Review E*, (2003), 67, 011910.

[30] S.E. Feller, K. Gawrisch, and A.D. MacKerell, *Journal of the American Chemical Society*, (2002), 124, 318-326.

[31] T. Huber, K. Rajamoorthi, V.F. Kurze, K. Beyer, and M.F. Brown, *Journal of the American Chemical Society*, (2002), 124, 298-309.

[32] A.L. Rabinovich, P.O. Ripatti, and N.K. Balabaev, *Russian Journal of Physical Chemistry*, (2004), 78, 1004-1009.

[33] W.L. Jorgensen, J. Chandrasekar, J.D. Madura, R.W. Impey, and M.L. Klein, *The Journal of Chemical Physics*, (1983), 79, 926-935.

[34] W.D. Cornell, P. Cieplak, C.L. Bayly, I.R. Gould, K.M. Merz, D.M. Ferguson, D.C. Spellmeyer, T. Fox, J.W. Caldwell, and P.A. Kollman, *Journal of the American Chemical Society*, (1995), 117, 5179-5197.

[35] T.R. Stouch, K.B. Ward, A. Altieri, and A.T. Hagler, *Journal of Computational Chemistry*, (1991), 12, 1033-1046.

[36] P.J. Flory, *Statistical mechanics of chain molecules*, Interscience, New York, 1969.

[37] V.G. Dashevskii, *Conformational analysis of organic molecules*, Nauka, Moscow, 1982 (*in Russian*).

[38] S.E. Feller, D. Yin, R.W. Pastor, and A.D. Mackerell, *Biophysical Journal*, (1997), 73, 2269-2279.

[39] M. Schlenkrich, J. Brickmann, A.D. MacKerell, and M. Karplus, *In: "Biological Membranes"*, K. Merz and B. Roux eds., Birkhauser, Boston, 1996, 31-81.

[40] N.K. Balabaev and A.A. Lemak, *Molecular Simulation*, (1995), 15, 223-231.

[41] R. Koynova and M. Caffrey, *Biochimica et Biophysica Acta*, (1998), 1376, 91-145.

[42] M.C. Wiener and S.H. White, *Biophysical Journal*, (1992), 61, 434-447.

[43] M.T. Hyvönen, T.T. Rantala, and M. Ala-Korpela, *Biophysical Journal*, (1997), 73, 2907-2923.

[44] A.L. Rabinovich, P.O. Ripatti, and N.K. Balabaev, *Journal of Biological Physics*, (1999), 25, 245-262.

[45] A.L. Rabinovich, P.O. Ripatti, and N.K. Balabaev, *Proceedings of SPIE*, (1999), 3687, 175-181.

[46] A.L. Rabinovich, P.O. Ripatti, and N.K. Balabaev, *Proceedings of SPIE*, (2000), 4064, 144-155.

[47] A.L. Rabinovich, P.O. Ripatti, and N.K. Balabaev, *Russian Journal of Physical Chemistry*, (2000), 74, 1809-1815.

[48] V.V. Kornilov, A.L. Rabinovich, and N.K. Balabaev, *Proceedings of SPIE*, (2002), 4627, 129-140.

[49] V.V. Kornilov, A.L. Rabinovich, and N.K. Balabaev, *Russian Journal of Physical Chemistry*, (2002), 76, 1819-1823.

[50] V.V. Kornilov, A.L. Rabinovich, and N.K. Balabaev, *Russian Journal of Physical Chemistry*, (2004), 78, 1010-1013.

[51] A.L. Rabinovich and N.K. Balabaev, *Proceedings of SPIE*, (2001), 4348, 215-224.

[52] L. Saiz and M.L. Klein, *Biophysical Journal*, (2001), 81, 204-216.

[53] A.L. Rabinovich and P.O. Ripatti, *Biophysics*, (1997), 42, 879-887.

[54] A.L. Rabinovich and P.O. Ripatti, *Biophysics*, (1998), 43, 268-275.

[55] A.L. Rabinovich and P.O. Ripatti, *Russian Journal of Physical Chemistry*, (1998), 72, 590-594.

[56] A.L. Rabinovich and P.O. Ripatti, *Proceedings of SPIE*, (1998), 3345, 193-197.

[57] A.L. Rabinovich and P.O. Ripatti, *Membrane Cell Biology*, (2000), 13, 697-714.

[58] A.L. Rabinovich and P.O. Ripatti, *Biophysics*, (2000), 45, 790-796.

[59] A.L. Rabinovich and P.O. Ripatti, *Biophysics*, (2000), 45, 797-804.

[60] A.L. Rabinovich, P.O. Ripatti, and N.K. Balabaev, *Russian Journal of Physical Chemistry*, (2002), 76, 1824-1828.

[61] H. Heller, M. Schaefer, and K. Schulten, *The Journal of Physical Chemistry*, (1993), 97, 8343-8360.

In: Molecular Simulation Studies in Material ...
Editor: Kholmirzo Kholmurodov, pp. 31-39
ISBN 1-59454-607-X

Chapter 4

COMPUTATIONAL MEMBRANOLOGY: MOLECULAR DYNAMIC STUDY OF PHOSPHOLIPID MEMBRANES

A. A. Shahinyan and K. A. Shahinyanet
The Institute of Applied Problems of Physics
National Academy of Sciences of Armenia
Hrachya Nersessian str. 25, 375014 Yerevan, Republic of Armenia
A. H. Poghosyan and G. A. Yeghiazaryan
The International Scientific-Educational Center
National Academy of Sciences of Armenia
Marshall Baghramian Ave. 24d, 375019 Yerevan, Republic of Armenia

Abstract

Some of the structural and functional processes taking place in the mono- and multi-lipid bilayers are investigated at time scales up to 10 ns using the methods of computer simulations. We report on the development of a special code "MDesigner" intended for construction of lipids and lipid bilayers.

1. Introduction

Phospholipid bilayers have been investigated for many years as a structural model for biological membranes [1]. Dipalmitoylphosphatidylcholine (DPPC) has been regarded as the benchmark lipid [2, 3], since many experiments were performed specially with this lipid. A lot of experiments [4, 5, 6] have been carried out to study the phospholipid-water interface, processes of hydration and water molecules orientation on the boundary of the phospholipid bilayer. During the last years a number of papers have been devoted to the molecular dynamic simulation (MD) of those structures [7, 8, 9, 10].

In present paper, using the methods of MD simulation, we study several physical and chemical properties of phospholipid bilayers, viz.:

[i] the change of the average volume of the hydrophobic and hydrophilic parts in the total volume of the bilayer and variation of their ratio within the time interval of 3-10ns;

[ii] the distribution of charged -N(CH3) 3+ parts of choline polar groups of the phospholipid molecules on the surface and the terminal methyl groups -CH3 of phospholipid hydrocarbon chains in hydrophobic core of bilayer;

[iii] the structure of multi-lipid bilayer.

2. Computational Details

We studied a phospholipid bilayer and water system consisting of 128 DPPC molecules and 3655 water molecules in the case of mono-lipid bilayer, and 64 DPPC and 64 Dimiristoilphosphatidylcholine (DMPC) molecules (the ratio of two types of phospholipids is DMPC: DPPC=1:1) and 4224 water molecules in case of the multi-lipid bilayer. The initial system comprised 17365 atoms. Atomic charges were taken from quantum mechanical calculation [11, 12]. The lipid force field parameters were taken from [13]. The SPC parameters of water molecules were taken from [14]. After the creation of initial configuration of bilayer, the water was added with the thickness of on each side of bilayer and some unwanted water molecules from hydrohobic region were removed. The initial configuration of bilayer was stabilized in 5000 steps by Newton-Raphson energy minimization algorithm. The MD simulations was done for the time interval 3-10ns. In computer experiment the temperature and pressure were set to 325K and 1 atm, respectively, both of them being controlled by the Berendsen coupling method. The temperature and the pressure both were maintained using Langevin dynamics and Nose-Hoover Langevin piston methods, respectively. The SHAKE algorithm for constraining lengths of the bonds. The simulation time step of 2fs was used with the list of non-bonded pairs being updates each 20fs. The long-range Coulomb and the short-range van der Waals interactions were cut-off's we set to 20Åand 14Å, respectively. All the hydrocarbon chains were set in all-trans configuration.

MD simulations and energy minimizations were performed using NAMD code with the CHARMM27 all-atom force field model, modified for lipids. The simulation was done on Pentium4 (2400MHz) and on Armcluster (National Academy of Sciences of Armenia) with the following parameters: Nodes/Processors: 64/128, Processor type - Intel Xeon 3.06 GHz, Memory - 64 * 1 Gb = 64 Gb, HDD - 64 * 36.7 Gb = 2348.8 Gb, Interconnect type-Myrinet, Managing Network type-Gigabit Ethernet, Node dimension - 2U, Peak Performance - 783.36 Gflops, OS - RedHat 9.0. During the simulation the coordinates were saved every 5ps.

The surface tension of the bilayer-water interface was set to 50.

3. Results and Discussions

For construction of lipids and lipid bilayers a new code "MDesigner" was developed. The present version of the code has an ability of constructing lipid molecules and lipid bilayers, exporting of the created structures in standard PDB format to be used by widely distributed MD simulation packages (GROMOS, NAMD). The program has an intuitively-clear graphic interface and the possibility of 3D visualization.

The main features of the program are:

- Import/Export of PDB Files
- 3D Visualization (Points, Balls and Cylinders)
- Translation, Rotation, Zoom
- Lipid Builder
- Bilayer Builder
- Replicator

The basic tools included in the program for work with lipids and lipid bilayers are the Lipid Builder, the Bilayer Builder and the Replicator.

The Lipid Builder (Fig.1) is used for creation of lipid molecules. The user can specify such parameters of lipid molecules as the polar head group, a choice of chemical bond lengths, and also torsion and valent angles. The given approach allows not only for building models of natural lipids, but also for the synthetic lipids with hydrocarbon chains of requested length.

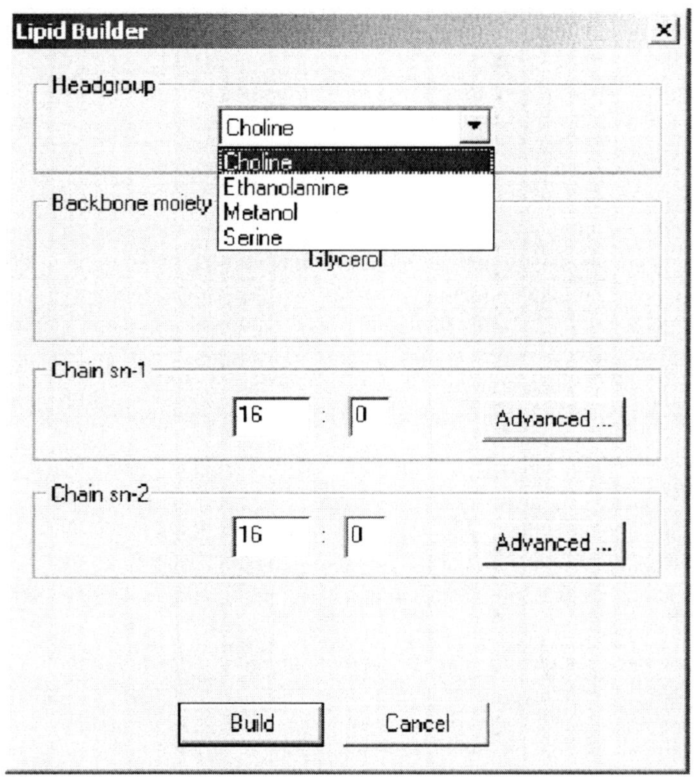

Figure 1. Lipid Builder dialog window

Bilayer Builder (Fig.2) is intended for creation of lipid bilayers on the base of already created lipids. There is also a possibility to creation of mixed bilayers with different lipids. For bilayer creation it is necessary to specify all lipid molecules to be incorporated in bilayer. For this there are buttons Add, Edit and Delete.

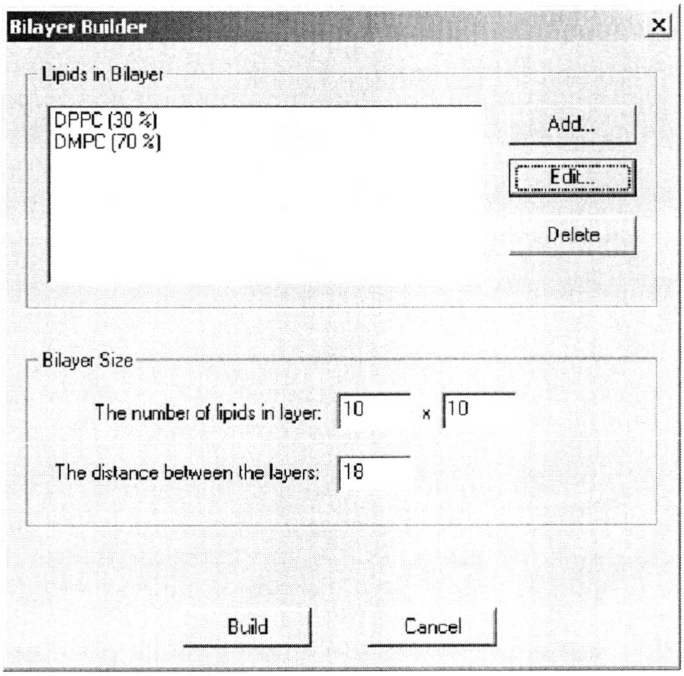

Figure 2. The Bilayer Builder dialog window.

For a new lipid addition in dialog window Add Lipid (Fig.3) it is necessary to specify a name of new lipid, a path to PDB file, probably created with Lipid Builder, the box size and the portion of the given type of lipid in total lipids in percentage.

In Fig.4 the dependence on the time of the ratio of average volumes of hydrophilic and hydrophobic parts of mono-lipid bilayer (Vphil/Vphob), consisting of from DPPC is shown. As it can be seen from the figure, Vphil/Vphob changes from 0,24 to 0,32.

The observed fluctuations are conditioned on thermal movements of molecules in the system.

Taking into account, that the interaction of other molecules of the system with the phospholipid bilayer must depend of on their hydrophilic and hydrophobic properties, and should get to their localization in the hydrophilic or hydrophobic domains of the bilayer,with different dielectric constants, we may consider that in the above case the interacting molecules may fall under the influence of the different electrostatic fields of the charged surfaces of bilayer.

On the Fig.5 the dynamics of displacements of water molecules, taken of from different levels of polar value of bilayer are represented.

Displacement of the molecules of water from three different levels: a) from the lower (near the glycerol groups); b) middle (near the choline groups) levels of polar part of bilayer and c) bulk (tightly bound and free) water are examined. As we can see from figure, the jumping of water molecules from more bounded state to a less one can take place.

Naturally, the state and the position of water and other molecules in bilayer must depend on the spatial distribution and the displacements of charged and hydrophobic hydrocarbon groups of phospholipid molecules, correspondingly in the hydrophilic and hydrophobic

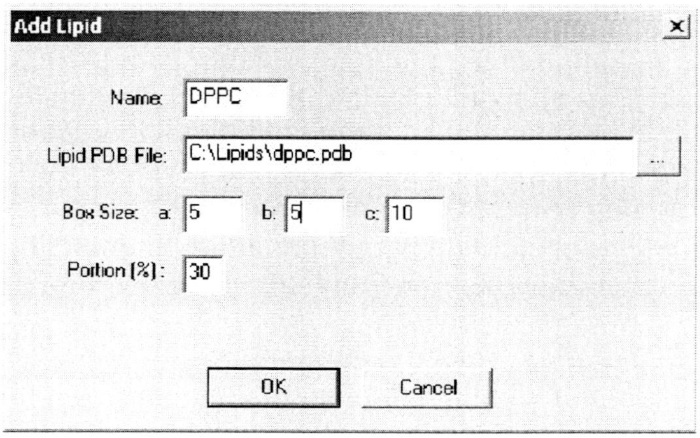

Figure 3. Add Lipid dialog window

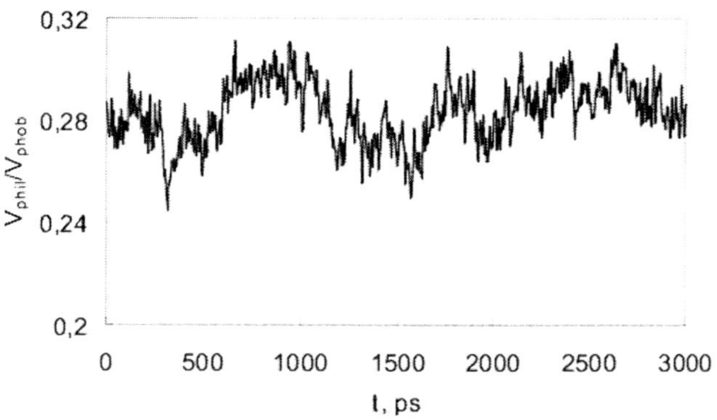

Figure 4. The dependence on the time of the ratio of average volumes of hydrophilic and hydrophobic parts of phospholipid (DPPC) bilayer.

parts of bilayer. Therefore, we simulate the distribution of polar –N(CH3)3+ and terminal -CH3 groups of phospholipid molecules, correspondingly on the surface and in the hydrocarbon core of bilayer, by the method of MD.

From the spatial and flat patterns of mentioned groups distribution (Fig.6-8) we can infere that:

[a] bilayer has structurally non-uniform hydrophilic surface and hydrophobic core;

[b] on the surface of the bilayer the "cavities" and "heights" are observed, at that the sizes of "cavities" are comparable with the sizes of great number of molecules or groups of molecules;

[c] within these "cavities" the drastic change of the potential of the electric field of charged groups of the phospholipid molecules, as compared to the surrounding, should be observed, that in its turn must affect the chemical and physical processes taking place in the molecules located in these "cavities".

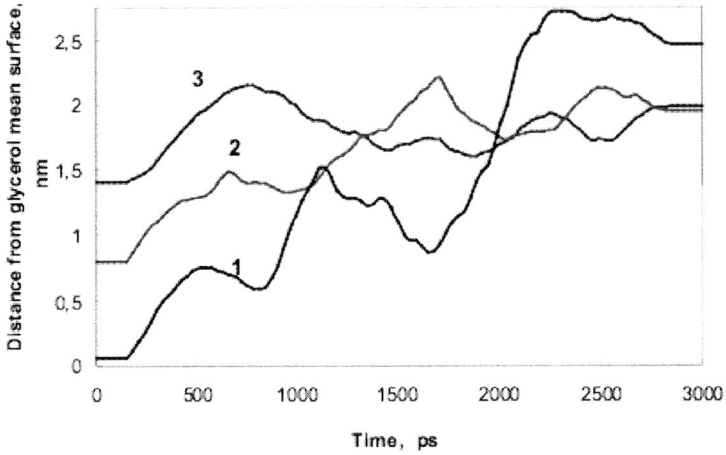

Figure 5. Time dependence of vertical, to the surface of bilayer, displacement of random molecules of water from different levels of polar value of bilayer: 1 – near the glycerol groups; 2 – near the choline groups; 3–bulk water

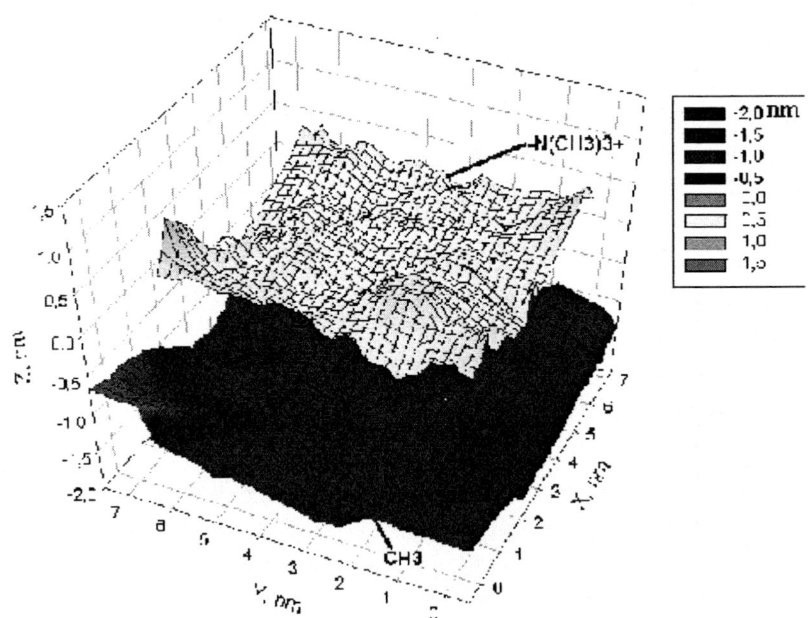

Figure 6. The 3D distribution of positively charged amino groups (yellow) and the terminal CH3 groups (blue) of hydrocarbon chains of phospholipid molecules. The simulation time is 3 ns

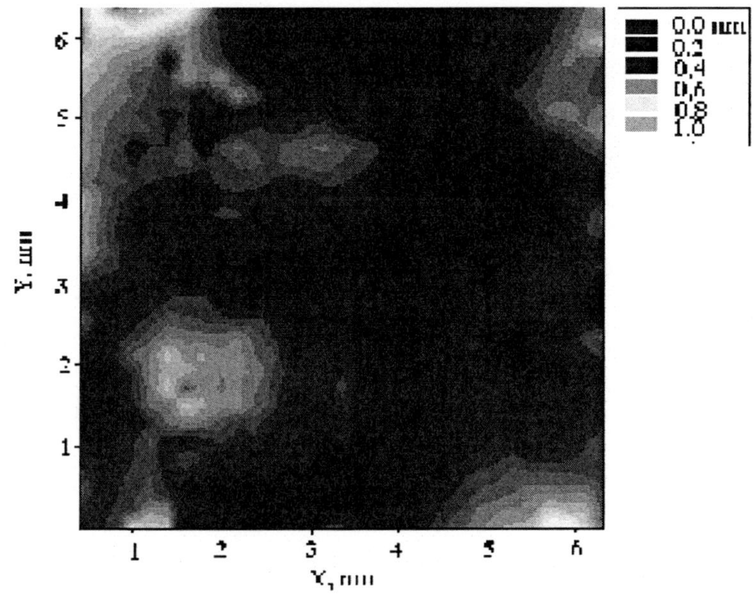

Figure 7. The distribution of positively charged amino groups of phospholipid molecules (plane view) of the bilayer.

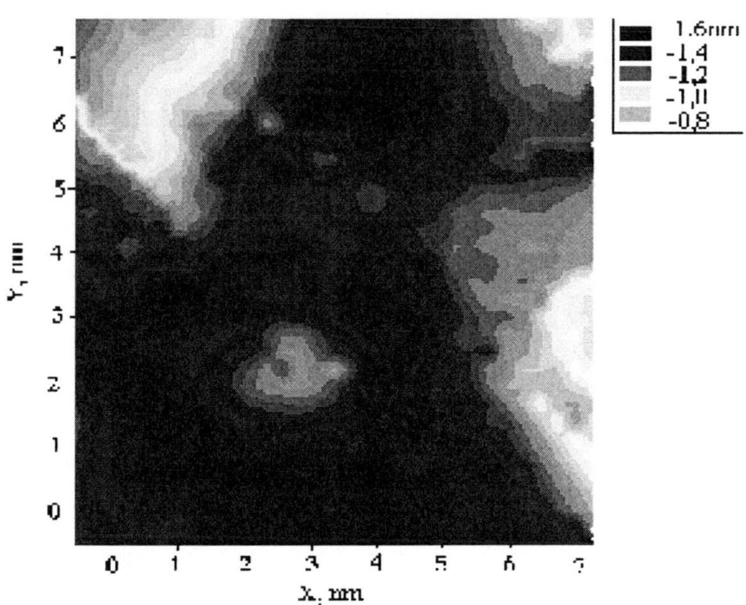

Figure 8. The distribution of the terminal CH3 groups of hydrocarbon chains of phospholipid molecules (plane view) in the hydrophobic volume of the bilayer

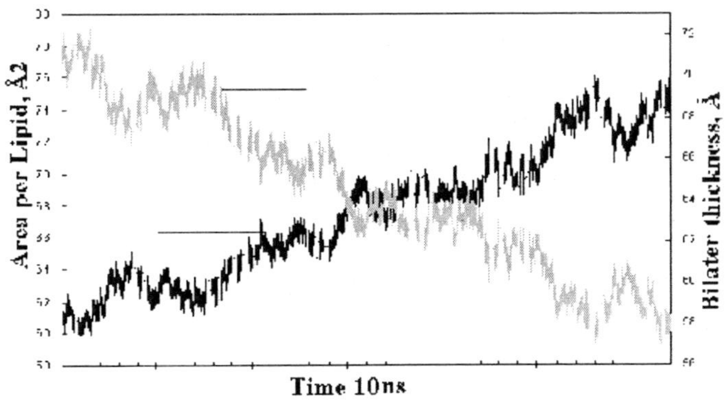

Figure 9. The area per lipid and Bilayer thickness against the time for multi-lipid bilayer consisting of from 50%: 50% of DPPC and DMPC

Some interesting results are obtained for multi-lipid bilayer, consisting of 50% DPPC and 50% DMPC. In Fig.9the dynamics of changing of multi-lipid bilayer thickness and the area per lipid on the bilayer surface are shown .

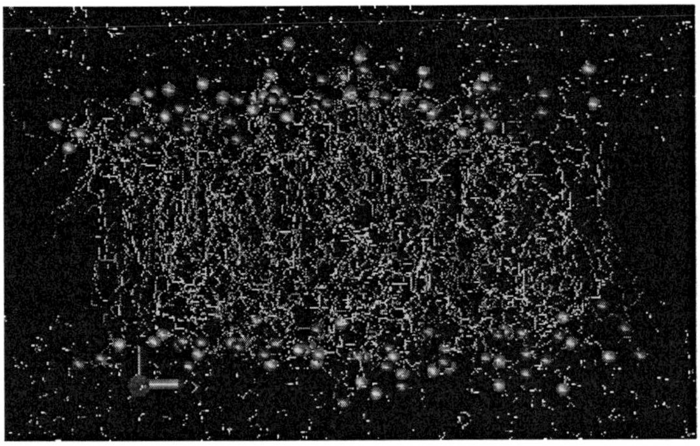

Figure 10. Multi-lipid Bilayer consisting of from 50%: 50% of DPPC (green) and DMPC (red). After 10 ns simulation.

As we see from figures, the addition of the DMPC to the DPPC bilayer leads to the decrease of bilayer's thickness and to the increase of area per lipid on the bilayer surface. As it is observed from Fig.10, the mutual penetration of hydrocarbon chains of two opposite sides of bilayer takes place. The possible reason for this kind of transformation of bilayer structure maybe the disturbance of balance of hydrophilic and hydrophobic interaction of lipid molecules in bilayer caused by replacement of phospholipid molecules with longer hydrocarbon chains by shorter ones. Thus, some of the pure structural and functional processes taking place in the mono-lipid and multi-lipid bilayers at the time interval up to 10 ns were investigated by the method of computer molecular simulation.

Acknowledgments

The work was supported in part by the International Science and Technology Center (ISTC) under grant "ARMCLUSTER" (*№* A-823).

References

[1] L. Albert Leninger, The Molecular Basis of Cell Structure and Function, *Biochemistry*, Worth Publishers, INC. NY, 1972.

[2] J. Nagle, et al., *Biophys. J.*, (1996), 70, 1419-1431.

[3] O. Berger, et al., *Biophys. J.*, (1997), 72, 2002-2013.

[4] E.G. Finer and A. Darke, *Chem Phys. Lipids.*, (1974), 12, 1-16.

[5] R.P. Rand and V.A. Parsegian, *Biochim. Biophys. Acta.*, (1989), 988, 351-376.

[6] W. Hubner and A. Blume, *Chem. Phys. Lipids.*, (1998), 96, 99-123.

[7] H. Alper, et al., *J. Chem. Phys.*, (1993), 98, 9798-9807.

[8] S.J. Marrink, et al., *Langmuir.*, (1993), 9, 3122-3131.

[9] K.V. Damodaran and K.M. Merz, *Biophys. J.*, (1994), 66, 1076-1087.

[10] A.A. Shahinyan, et al., *Int. J .Mod. Phys.*, (2000), 11(1), 153-156.

[11] S.W. Chiu, et al., *Biophys. J.*, (1995), 69, 1230-1245.

[12] D.P. Tieleman and H.J.C. Berendsen, *J. Chem. Phys.*, (1996), 105, 4871-4880.

[13] O. Berger, et al., *Biophys. J.*, (1997), 72, 2002-2013.

[14] H.J.C. Berendsen, et al., *In: "Intermolecular Forces. Reidel Pub. Comp. Dordrecht"*, (1981), 331-342.

In: Molecular Simulation Studies in Material ... ISBN 1-59454-607-X
Editor: Kholmirzo Kholmurodov, pp. 41-52

Chapter 5

EFFECTS OF THE TEMPERATURE AND UNSATURATION ON THE STRUCTURE AND DYNAMICS PROPERTIES OF UNSATURATED LIPID MONOLAYERS: MOLECULAR DYNAMICS SIMULATIONS

***V. V. Kornilov*[*] *and N. K. Balabaev*[†]**
Institute of Mathematical Problems of Biology, Russian Academy of Sciences, Pushchino, Moscow Region, 142290, Russia
***A. L. Rabinovich*[‡]**
Institute of Biology, Karelian Research Centre, Russian Academy of Sciences, 11 Pushkinskaya str., Petrozavodsk 185910, Russia

Abstract

Molecular dynamics (MD) simulations were performed for six model monolayers of diacylglycerolipids (DG) with different number of cis double $C = C$ bonds of the sn-2 hydrocarbon chains. The simulation monolayers consisted of 48 identical lipid molecules arranged in rectangular simulation cells, with periodic boundary conditions in the surface plane (X, Y directions). The temperature (T), the number of atoms (N) and the surface area per molecule (A) were fixed; two series of the MD simulations were performed under two different (fixed) temperatures. The simulation parameters have provided a liquid crystalline state of the monolayers, the MD trajectories were 5000 ps. The effect of the degree of unsaturation of the sn-2 hydrocarbon chains as well as the temperature influence on the lipid properties have been studied. The results of the simulations are compared with the available experimental data and other simulations of similar systems.

---

[*]E-mail address: balabaev@impb.psn.ru
[†]E-mail address: basil@psn.ru
[‡]E-mail address: rabinov@krc.karelia.ru

1. Introduction

The considerable quantity of fatty acyl chains of lipid molecules of most biological membranes contains carbon-carbon double bonds [1]-[3]. It is widely recognized that these unsaturated (with one $C=C$ bond) and especially polyunsaturated (with three or more, up to six $C=C$ bonds) lipid molecules play an essential role in different membrane processes, but it is not exactly known how lipid unsaturation participates in different membrane functions and membrane organization [4]-[15]. In the last years lipid bilayer and monolayer systems have been extensively studied as model systems for biological membranes using the computer simulation methods [16]-[20].

In this paper the results of molecular dynamics (MD) simulations of six model monolayers of diacylglycerolipids (DG) with different number of cis double $C = C$ bonds in the *sn*-2 hydrocarbon chain have been discussed. One monolayer consisted of saturated DG molecules 18:0/18:0 (1,2-stearoyl-*sn*-3-glycerolipids) and the molecules of the other monolayers contain *sn*-1-saturated, *sn*-2-unsaturated DGs: 18:0/18:1(n-9)cis DG, 1-stearoyl-2-oleoyl-*sn*-3-glycerolipid; 18:0/18:2(n-6)cis DG, 1-stearo-yl-2-linoleoyl-*sn*-3-gly-cerolipid; 18:0/18:3(n-3)cis DG, 1-stearoyl-2-lino-lenoyl-*sn*-3-glycerolipid; 18:0/20:4(n-6)cis DG, 1-stearoyl-2-arachidonoyl-*sn*-3-glycerolipid; 18:0/22:6(n-3)cis DG, 1-stearoyl-2-docosahexaenoyl-*sn*-3-glycerolipid. In the notation m:x(n-y) the m is the total number of carbon atoms in the chain, x is the number of methylene-interrupted double bonds and y denotes the number of carbons between the final CH_3 group of the chain and the nearest double bond.

2. Model

In our model the lipid hydrocarbon chains were treated in accordance with their real chemical structure and the head groups of the molecules were approximated by united atoms [21]. The simulation systems consisted of 48 identical lipid molecules arranged in a rectangular box, with periodic boundary conditions in the surface plane (X, Y directions). A typical snapshot for one of the investigated monolayers along the MD trajectory is presented in Fig.1.

All the atoms in the monolayer systems were treated classically. The total potential energy U of the system included the sum of the intramolecular energy of the lipid molecules U_{intra}, the intermolecular energy U_{inter}, and the surface interaction energy U_{surf}:

$$U = \sum U_{intra} + U_{inter} + \sum U_{surf}. \tag{1}$$

The potential energy of the molecule U_{intra} was calculated as the sum of the energy of nonbonded (van der Waals) interactions, the bond-stretching energy, the angle-bending energy, the torsion motion energy, and the out-of-plane energy:

$$U_{intra} = \sum\sum U_{LJ}(r_{ij}) \cdot W_{nb}(r_{ij}) + \sum K_l(l-l_0)^2 + \\ \sum K_\theta(\theta-\theta_0)^2 + \sum K_\varphi\left[1+\delta\cos(n_0\varphi)\right] + \sum K_\phi\left[1-\cos(2\phi)\right] \tag{2}$$

where

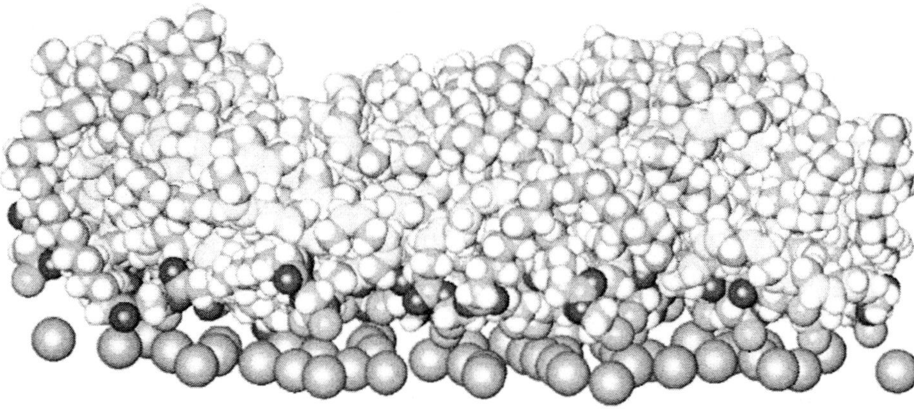

Figure 1. A typical snapshot for 18:0/22:6 DG monolayer (48 lipid molecules in the calculation box) along the MD trajectory. Color coding: carbon atoms C are green, carbon atoms of double bonds $C = C$ are yellow, hydrogens H are grey, oxygens O - red, the united head group atoms are cyan

$$U_{LJ}(r_{ij}) = 4\varepsilon_{ij}[(\sigma_{ij}/r_{ij})^{12} - (\sigma_{ij}/r_{ij})^{6}], \quad (3)$$

$$W_{nb}(r_{ij}) = \begin{cases} 1, & r_{ij} \le R_{on} \\ \frac{(R_{off}^2 - r_{ij}^2)^2 (R_{off}^2 - 3R_{on}^2 + 2r_{ij}^2)}{(R_{off}^2 - R_{on}^2)^3}, & R_{on} < r_{ij} < R_{off} \\ 0, & r_{ij} \ge R_{off} \end{cases} \quad (4)$$

If i and/or j correspond to the united head atom of the molecule, then

$$U_{nb} = U_{LJ}(r_{ij}) + \varepsilon_{ij}, \quad (5)$$

at $r_{ij} \le 2^{1/6} \cdot \sigma_{ij}$;

$$U_{nb} = 0, \quad (6)$$

at $r_{ij} > 2^{1/6} \cdot \sigma_{ij}$.

Here r_{ij} is the distance between atoms i and j, ε_{ij} and σ_{ij} are Lennard-Jones parameters, l is the bond length, θ is the bond angle, ϕ is the torsion angle, ϕ - the out-of-plane angle, n_0 is the dihedral multiplicity, $\delta = \pm 1$; l_0 , θ_0 are the equilibrium values for the bond lengths and angles; K_l , K_θ , K_ϕ and K_ϕ are force constants. For the van der Waals interactions the parameters of Lennard-Jones potentials were borrowed from [22], the other parameters - from [23].

The intermolecular energy U_{inter} was calculated as the sum of energies of nonbonded interactions $U_{LJ}(r_{ij}) \cdot W_{nb}(r_{ij})$ over all the intermolecular pairs of atoms. The surface interaction energy U_{surf} was taken as the sum of the "hydrophilic" interactions U_{h_phil} of the lipid head units ("atoms"), and "hydrophobic" interactions $\sum U_{h\_phob}$ of all the other atoms with the monolayer interface (which was modeled by a flat surface):

$$U_{h\_phil} = (1/2) \cdot C_{surf} \cdot (Z - Z_{surf})^2, \quad (7)$$

$$U_{h\_phob} = C_{srb}[(R_{srb}/Z)^9 - 3(R_{srb}/Z)^3 + 2], \quad (8)$$

when $0 < Z \leq R_{srb}$, and

$$U_{h\_phob} = 0, \quad (9)$$

when $Z < 0, Z > R_{srb}$. Here C_{srb} , R_{srb} , C_{surf} are the parameters [21, 23].

All the monolayers were studied under constant surface area per molecule (*NAT* ensemble: the number of atoms N, the surface area per molecule A and the temperature T were fixed). Two series of the MD simulations were performed under two different temperatures: for the five unsaturated DG monolayers the temperatures were 283 K and 303 K, and the temperatures for 18:0/18:0 DG monolayer were 310 and 330 K. The areas per lipid were: 0.654 nm^2 for monolayer 18:0/18:0 DG, 0.666 nm^2 for 18:0/18:1 DG, 0.673 nm^2 for 18:0/18:2 DG, 0.666 nm^2 for 18:0/18:3 DG, 0.706 nm^2 for 18:0/20:4 DG and 0.716 nm^2 for 18:0/22:6 DG monolayer, as in the experimental work [24] for phosphatidylcholines with the same fatty acid contents. The conditions of all the MD computer simulations have provided a liquid crystalline state of the monolayers.

The movement of all atoms in the system was described by Newton's equations of motion, Verlet's algorithm [25] was used for their integration, and Berendsen's thermostat [26] was used for temperature control:

$$m_i \frac{d\vec{v}_i}{dt} = -\frac{\partial U}{\partial \vec{r}_i} + \beta_T \cdot \left(\frac{T_{ref}}{T} - 1 \right) \cdot \vec{v}_i \quad (10)$$

where i = 1, 2, ..., N; N is the number of atoms, m_i is the mass of atom i; $\vec{r}_i$, $\vec{v}_i$ are the coordinates and velocities of atom i; $\vec{\chi}$, β_T are the parameters for the temperature T relaxation, T_{ref} is prescribed temperature. The simulation time step was 1 fs. The MD trajectories of all the monolayers were 5 ns.

3. Results

3.1. Order Parameters

Bond order parameter profiles are of considerable importance in defining membrane properties. The lipid hydrocarbon chain C-H bond order parameters (Fig.2) were calculated in this work as

$$S_{CH} = (1/2) \cdot (3 < \cos^2 \beta_{CH} > -1), \quad (11)$$

where β_{CH} is the angle between the C-H bond and the monolayer normal. The -S_{CH} profiles of sn-1 chains of all the monolayers proved to be qualitatively similar to those obtained in experimental studies. Indeed, there is a plateau region for the several CH_2-groups next to the head group, then the -S_{CH} values drop towards zero [27]-[29].

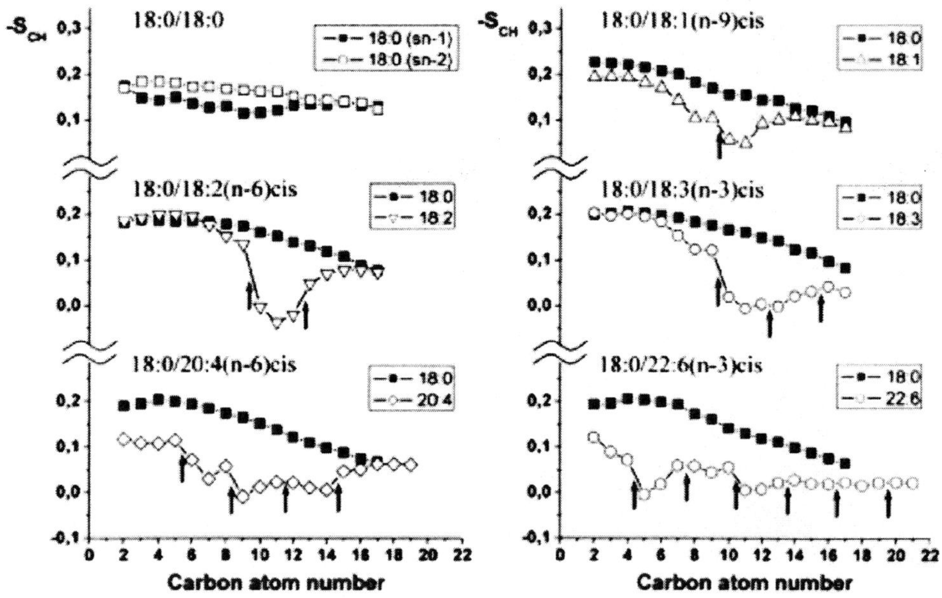

Figure 2. Lipid hydrocarbon chain order parameter $-S_{CH}$ profiles for all the monolayers investigated. The arrows indicate the double bond positions. Carbon atom No.1 is the carbon of $C = O$ group of the hydrocarbon chain

The bond order parameter profiles for all the unsaturated chains proved to be significantly differ from those of the saturated ones: the dips in the values are observed near the locations of cis double $C = C$ bonds. Similar deuterium order parameter profiles $|S_{CD}|$ are known from the experiments [30]-[35] for 18:1, 18:2 and 22:6 chains. These results are also in good agreement with the results of our MD simulation of the same monolayer under different conditions [36, 37], and with the MD studies of bilayers of unsaturated lipids [38]-[40].

3.2. Distribution Functions

The detailed description of the ordering picture of the C-H vectors can be obtained from the bond-vector orientation distributions $\rho_{CH}(\beta_{CH})$, where β_{CH} are the angles between the C-H bond and the normal. The relation of the bond orientation distributions $\rho(\beta)$ to the order parameter S were analyzed in terms of angles β^{max} corresponding to the maximum of distribution functions, and widths $\delta\beta$ of the distributions $\rho(\beta)$ corresponding to half-heights of the distributions [41]. It is clear that β^{max} is preferred orientation of the bond (i.e., a "geometric" factor of the bond ordering) and $\delta\beta$ is a factor of the bond orientation "fluctuations".

Fig.3 shows the values of the C-H bond-vector orientation distribution function widths $\delta\beta_{CH}$ corresponding to half-heights of the distributions $\rho_{CH}(\beta_{CH})$ of hydrocarbon chains of all investigated DG monolayers. The values of $\delta\beta_{CH}$ for methylene groups located between CH=CH groups are rather high than that for C-H bonds of double-bonded carbons. This means that there are significant orientation fluctuations of C-H bonds of the carbons located

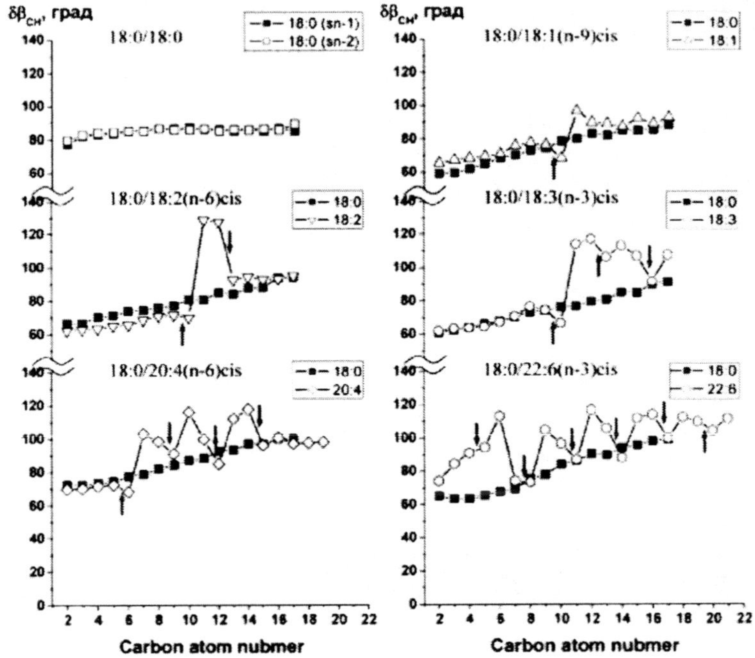

Figure 3. Values of $C-H$ bond-vector orientation distribution functions widths $\delta\beta_{CH}$ corresponding to half-heights of the distributions $\rho_{CH}(\beta_{CH})$ for all the hydrocarbon chains of DG monolayers. The arrows indicate the double bond positions. Carbon atom No.1 is the carbon of $C=O$ group of the hydrocarbon chain

between double-bonded carbons. This effect is consistent with the results of Monte Carlo simulations of isolated polyunsaturated hydrocarbon chains [41] and previous MD studies of DG monolayers [21, 36, 37]. One of the main results from the analysis of the data for the distributions $\rho(\beta)$ investigated is that the fluctuation factors $\delta\beta$ depend on both the segment location in the chain and the segment chemical structure. Moreover, it should be mentioned, that these results can explain the high flexibility of unsaturated chains and allow understanding the mechanism of increasing the membrane fluidity with the increase of the degree of unsaturation of lipids.

3.3. Diffusion of Lipid Molecules

In Fig.4 the time dependencies of the mean-square-displacement (MSD) of the center of mass of lipid molecules are presented. It is seen that the diffusion motion is increase with increasing the number of C=C bonds in the sn-2 chain of the DG molecule. From the lateral displacement of the center of masses, we calculated the lateral diffusion coefficient. In principle, it can be obtained from MD simulations as the ratio of MSD $(\vec{r}(t)-\vec{r}(o))^2$ to time in the limit of infinite t:

$$D_{xy} = \lim_{\tau\to\infty} \frac{\left\langle |\vec{r}(t_0+\tau)-\vec{r}(t_0)|^2 \right\rangle}{2N_f\tau} \qquad (12)$$

The brackets indicate an average over different times t_0 and all the atoms. The parameter N_f describes the number of translation degrees of freedom. For the lateral diffusion of lipids N_f is 2.

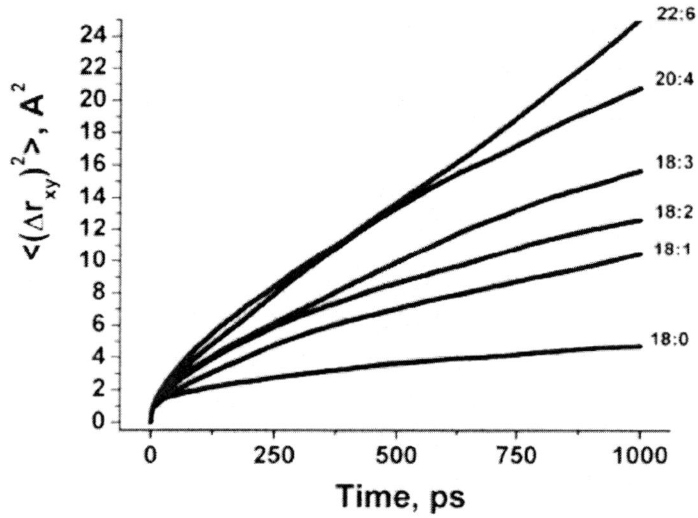

Figure 4. The mean-square-displacement of the center of mass of lipid molecules for all the monolayers investigated

The diffusion coefficient is found to increase with increasing chain unsaturation. The diffusion coefficient of the lipids in the monolayer 18:0/18:0 DG is min, D_{xy}=0,3·10^{-8} cm^2/s, and in the monolayer 18:0/22:6 DG it is max, D_{xy}=2,6·10^{-8} cm^2/s. Our values of the diffusion coefficients are in reasonable agreement with the values obtained from the neutron scattering experiments [42]-[44].

3.4. The Temperature Coefficients of the Order Parameter

The temperature coefficients of the order parameter of C-H bonds are presented in Fig.5. The temperature dependencies of S_{CH} are different for different segments of the chain. In the 18:0 hydrocarbon chains, the part which is nearest to the head of the lipid molecule is the most sensitive to the temperature. It should be mentioned that the derivative dS_{CH}/dT exhibits wider variations along the unsaturated chains as compared to the 18:0 chain. It is seen that the temperature sensitivity of C-H bond order parameter is significantly dependent on the lipid unsaturation and location of double bonds. The value of dS_{CH}/dT for the saturated chain 18:0 agrees qualitatively with the experimental data reported in [45]; to our knowledge there is no experimental data on dS_{CH}/dT for unsaturated chains.

3.5. Conclusions

Thus, the results of these MD simulations are in a good qualitative agreement with the available experimental data and with other computer simulations of similar systems. Accordingly, a "united atom" (i.e., coarse grained) approximation of lipid head groups in a

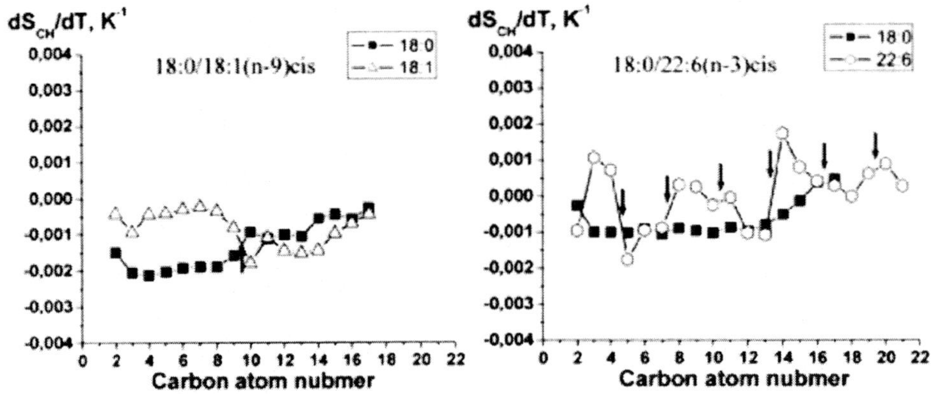

Figure 5. Temperature coefficients of the order parameter of C-H bond in the 18:0/18:1(n-9)cis DG and 18:0/22:6(n-3)cis DG monolayers. For notations and numeration of the atoms, see Fig.2

monolayer (like the head groups of the DG molecules used in the present work) is acceptable for studies of the characteristics of the lipid hydrocarbon chains in the system.

Both the bond order profile results and the results of the analysis of the bond orientation distributions can explain the high values of the flexibility of unsaturated chains and allow to understand the molecular mechanism of increasing the membrane fluidity with the increase of the degree of unsaturation of lipids. It is shown that temperature sensitivity of the DG molecule properties is greatly dependent on the number of C=C bonds in the sn-2 chain. Hydrocarbon chains of highly unsaturated lipids are characterized by higher mobility and lower temperature sensitivity. These properties can be used to explain the role of polyunsaturated lipids in the temperature adaptations of biological membranes.

Acknowledgments

This work has been supported by Russian Foundation for Basic Research (grant No. 03-03-32229) and INTAS (grant No. 01-0067).

References

[1] R.B. Gennis, *Biomembranes: molecular structure and function* , New York: Springer-Verlag, 1989.

[2] A.L. Lehniger, D.J. Nelson, and M.M. Cox, *Principles of biochemistry*, New York, Worth, 1993.

[3] P.L. Yeagle, *The membrane of cells*, Academic Press, San Diego, CA, 1993.

[4] D. Chiras, *Human biology*, 3rd ed. by C. Starr, and B. McMillan, Wadsworth Publishing, 1999.

[5] D.J. Murphy, The molecular organization of the photosynthetic membranes of higher plants, *Biochim. Biophys. Acta*, (1986), 864, 33-94.

[6] A.E. Anderson, P.J. O'Brien, R.D. Wiegand, C.A. Koutz, A.M. Stinson, Conservation of docosahexaenoic acid in the retina., *In: Neurobiology of Essential Fatty Acids*, ed. by N.G. Bazan, et al., Plenum Press, New York, 1992, 295-306.

[7] D.C. Mitchell, K. Gawrisch, B.J. Litman, N. Salem Jr., Why is docosahexaenoic acid essential for nervous system function?, *Biochem Soc. Trans.*, (1998), 26, 365-370.

[8] N.G. Bazan, Supply of n-3 polyunsaturated fatty acids and their significance in the central nervous system, *In: Nutrition and the Brain*, Wurtman R. J., Wurtman J. J. eds., Raven Press New York, NY, 1990, 8, 1-24.

[9] J. Tinoco, Dietary requirements and functions of linolenic acid in animals, *Prog. Lipid Res.*, (1982), 21, 1-46.

[10] C.D. Stubbs, A.D. Smith, The modification of mammalian membrane polyunsaturated fatty acid composition in relation to membrane fluidity and function, *Biochim. Biophys. Acta*, (1984), 779, 89-137.

[11] R.R. Brenner, Effect of unsaturated fatty acids on membrane structure and enzyme kinetics, *Prog. Lipid Res.*, (1984), 23, 69-96.

[12] E.A. Dratz, A.J. Deese, The role of docosahexaenoic acid in biological membranes: examples from photoreceptors and model membrane bilayers, *In: Health Effects of Polyunsaturated Fatty Acids in Seafoods.*, ed. A.P.Simopoupous, R.R.Kiefer, and R.E.Martin, Academic Press, New York, 1986, 319-351.

[13] M.V. Bell, R.J. Henderson, J.R. Sargent, The role of polyunsaturated fatty acids in fish, *Comp. Biochem. Physiol. B.*, (1986), 83, 711.

[14] M. Neuringer, G.J. Anderson, W.E. Connor, The essentiality of n-3 fatty acids for the development and function of the retina and brain, *Annu. Rev. Nutr.*, (1988), 8, 517-541.

[15] N. Salem, H.-Y. Kim, J.A. Yergey, Docosahexaenoic acid: membrane function and metabolism, *In: Health Effects of Polyunsaturated Fatty Acids in Seafoods*, eds. A.P.Simopoupous, R.R.Kiefer, and R.E.Martin Academic Press, Orlando, 1986, 263-318.

[16] R.W. Pastor, Molecular dynamics and Monte Carlo simulations of lipid bilayers, *Current Opinion in Structural Biology*, (1994), 4, 486-492.

[17] D.P. Tieleman, S.J. Marrink, H.J.C. Berendsen, A computer perspective of membranes - molecular dynamics studies of lipid bilayer systems, *Biochimica et Biophysica Acta*, (1997), 1331, 235-270.

[18] K.M. Merz Jr., Molecular dynamics simulations of lipid bilayers, *Current Opinion in Structural Biology*, (1997), 7, 511-517.

[19] D.J. Tobias, K. Tu, M.L. Klein, Atomic-scale molecular dynamics simulations of lipid membranes, *Current Opinion in Colloid and Interface Science*, (1997), 2, 15-26.

[20] L. Saiz, M.L. Klein, Computer Simulation Studies of Model Biological Membranes, *Accounts of Chemical Research*, (2002), 35, 482-489.

[21] A.L. Rabinovich, P.O. Ripatti and N.K. Balabaev, Molecular dynamic investigation of bond ordering of unsaturated lipids in monolayers, *Journal of Biological Physics*, (1999), 25, 245-262.

[22] M. Schlenkrich, J. Brickmann, A.D. MacKerell, Jr., and M. Karplus, An empirical potential energy function for phospholipids: criteria for parameter optimization and applications, *In: Biological Membranes*, eds. K. Merz, Jr., and B. Roux, Birkhauser, Boston, 1996, 31-81.

[23] A.L. Rabinovich, P.O. Ripatti, and N.K. Balabaev, A molecular dynamics study of the characteristics of the molecules of polar diacylglycerolipids in layers, *Zhurnal fizicheskoi khimii*, (2000), 74, 1990-1996 (*in Russian*); *Russian Journal of Physical Chemistry* (*Engl. Transl.*), (2000), 74, 1809-1815.

[24] F. Separovic and K. Gawrisch, Effect of Unsaturation on the Chain Order of Phosphatidylcholines in a Dioleoylphosphatidylethanolamine Matrix, *Biophysical Journal*, (1996), 71, 274-282.

[25] L. Verlet, Computer "experiments" on classical fluids. I. Thermodynamical properties of Lennard-Jones molecules, *Physical Review*, (1967), 159, 98-114.

[26] H.J.C. Berendsen, J.P.M. Postma, W.F. van Gunsteren, A. DiNola, and J.R. Haak, Molecular dynamics with coupling to an external bath, *Chemical Physics*, (1984), 81, 3684-3689.

[27] J. Seelig, A. Seelig, Lipid conformation in model membranes and biological membranes, *Q. Rev. Biophys.*, (1980), 13, 19-61.

[28] J. Seelig, Deuterium magnetic resonance: Theory and application to lipid membranes, *Q. Rev. Biophys.*, (1977), 10, 353-418.

[29] M. Bloom , E. Evans, O.G. Mouritsen, Physical properties of the fluid lipid-bilayer component of cell membranes: a perspective, *Q. Rev. Biophys.*, (1991), 24, 293-397.

[30] J. Seelig, and N. Waespe-Šarcevic, Molecular Order in Cis and Trans Unsaturated Phospholipid Bilayers, *Biochemistry*, (1978), 17, 3310-3315.

[31] A. Seelig, and J. Seelig, Effect of a single cis double bond on the srtucture of a phospholipid bilayer, *Biochemistry*, (1977), 16, 45-50.

[32] H.U. Gally, G. Pluschke, P. Overath, and J.Seelig, Structure of E.coli membranes: Phospholipid conformation in model membranes and cells as studied by deuterium magnetic resonance, *Biochemistry*, (1979), 18, 5605-5610.

[33] J.E. Baenziger, H.C. Jarrell, R.J. Hill, and I.C.P. Smith, Average Structural and Motional Properties of a Diunsaturated Acyl Chain in a Lipid Bilayer: Effects of Two Cis-Unsaturated Double Bonds, *Biochemistry*, (1991), 30, 894-903.

[34] K. Rajamoorthi, and M.F. Brown, Bilayers of arachidonic acid containing phospholipids studied by ^{2}H and ^{31}P NMR spectroscopy, *Biochemistry*, (1991), 30, 4204-4212.

[35] T. Huber T., K. Rajamoorthi, V.F. Kurze, K. Beyer, and M.F. Brown, Structure of Docosahexaenoic Acid-Containing Phospholipid Bilayers as Studied by ^{2}H NMR and Molecular Dynamics Simulations, *Journal of the American Chemical Society*, (2002), 124, 298-309.

[36] V.V. Kornilov, A.L. Rabinovich, and N.K. Balabaev, Molecular Dynamics Analysis of the Structure and Properties of Monolayers of Polyunsaturated Lipids, *Zhurnal fizicheskoi khimii*, (2002), 76, 2018-2022 (*in Russian*); *Russian Journal of Physical Chemistry* (*Engl. Transl.*), (2002), 76, 1819-1823.

[37] V.V. Kornilov, A.L. Rabinovich and N.K. Balabaev, Molecular dynamics study of monolayers consisting of polyunsaturated diacylglycerolipids, *Proceedings of SPIE*, (2002), 4627, 129-140.

[38] A.L. Rabinovich, N.K. Balabaev, Molecular dynamics simulations of unsaturated lipid bilayers, *Proceedings of SPIE*, (2001), 4348, 215-224.

[39] L. Saiz, M.L. Klein, Influence of highly polyunsaturated lipid acyl chains of biomembranes on the NMR order parameters, *J. Am. Chem. Soc.*, (2001), 123, 7381-7387.

[40] T. Huber, K. Rajamoorthi, V.F. Kurze, K. Beyer, M.F. Brown, Structure of docosahexaenoic acid-containing phospholipid bilayers as studied by 2H NMR and molecular dynamics simulations, *J. Am. Chem. Soc.*, (2002), 124, 298-309.

[41] A.L. Rabinovich, and P.O. Ripatti, Computer simulations of hydrocarbon chains with cis-double bonds: An investigation of intramolecular bond order characteristics, *Proceedings of SPIE*, (1998), 3345, 193-197.

[42] W. Pfeiffer, T. Henkel, E. Sackman, W. Knoll, D. Richter, Local dynamics of lipid bilayers studied by incoherent quasi-elastic neutron scattering, *Europhys. Lett.*, (1989), 8, 201-206.

[43] S. Konig, W. Pfeiffer, T. Bayerl, D. Richter, E. Sackman, Molecular dynamics of lipid bilayers studied by incoherent quasi-elastic neutron scaterring, *J. Phys. II.*, (1992), 2, 1589-1615.

[44] E. Sackmann, Physical basis of self-organization and function of membranes: physics of vesicles, *In: Handbook of Biological Physics*, R. Lipowsky, and E. Sackmann eds., Elsevier, Amsterdam, 1995, 1A, 213-304.

[45] L.L. Holte, S.A. Peter, T.M. Sinnwell, Kl. Gawrish, 2H nuclear magnetic resonance order parameter profiles suggest a change of molecular shape for phosphatidylcholines containing a polyunsaturated acyl chain, *Biophys. J.*, (1995), 68, 2396-2403.

In: Molecular Simulation Studies in Material ... ISBN 1-59454-607-X
Editor: Kholmirzo Kholmurodov, pp. 53-66

Chapter 6

Chebyshev Polynomial Expansion Method for Nano-Quantum Systems

Toshiaki Iitaka* and Toshikazu Ebisuzaki
Computational Astrophysics Laboratory
RIKEN (The Institute of Physical and Chemical Research)
2-1 Hirosawa, Wako, Saitama 351-0198, Japan

Abstract

We review recent development of algorithm for linear response functions of quantum systems described by a very large Hamiltonian matrix. These numerical methods are combinations of numerical solution of the time-dependent Schroedinger equation, random vector representation of trace, and Chebyshev polynomial expansion of Fermi- and Boltzmann-operators, whose application includes optical absorption of nanocrystallites, and electron spin resonance (ESR) spectrum of one dimensional antiferromagnets at finite temperature.

1. Introduction

Computational physicists often face to problems of calculating the linear response functions or the real-time Green's functions of quantum manybody systems with degree of freedom $N \sim 10^6$ or more. Direct diagonalization of the Hamiltonian matrix is the most naive and powerful method for modest system size $N \leq 10^3$. However, it becomes prohibitive for larger system because the computational time grows as N^3. Therefore efficient numerical methods, such as Quantum Monte Carlo methods, Lanczos Methods, and Kernel Polynomial Methods have been developed and applied to various problems.

Quantum Monte Carlo methods (QMC)[1, 2] can generate the Green's functions of very large systems since QMC does not need to store the wave functions. They have been successfully used for evaluating the imaginary-time Green's functions and, therefore, various thermodynamic quantities. However, for evaluating dynamical quantities such as AC conductivity, one has to rely on numerical analytic continuation (e.g. *Maximum Entropy*

*E-mail address: tiitaka@riken.jp

Method [3]) to obtain the real-time Green's function from the imaginary-time one. This procedure is not straightforward because the statistical errors are amplified by numerical analytical continuation and the default model for the MEM must be assumed *a priori*.

Lanczos Methods (LM)[4, 5] have been useful techniques for evaluating dynamical responses of relatively large systems. The LM uses Lanczos recursion formula with Matrix Vector Multiplications (MVM), $|\phi'\rangle = H|\phi\rangle$, to tridiagonalize the Hamiltonian matrix, which leads to a continued fraction representation of the real-time Green's function. The drawback of LM is numerical instability for large numbers of MVM's, which originates in the algorithm. Recently, LM was extended to finite temperatures (Finite Temperature LM, FTLM) by introducing random sampling over the ground and excited eigenstates [6]. However, FTLM has a weak point that the number of excited eigenstates to be calculated increases rapidly as temperature or system size becomes large. Therefore reduction of computational costs by exploiting symmetries of the system becomes crucial.

Kernel Polynomial Method (KPM)[7, 8] calculates density of states and linear response functions by using Chebyshev polynomial expansion [9, 10, 11] of broadened delta functions. The Chebyshev polynomials are obtained through Chebyshev recursion formula by using MVM's. Unlike Lanczos recursion, Chebyshev recursion is free from the numerical instability observed in Lanczos recursion even for large numbers of MVM's. The use of KPM for linear response functions has been limited to the ground state calculation.

Time-Dependent method [12] is one of the most efficient and direct methods for calculating the real-time Green's functions and linear response functions, and has been successfully applied to electric conductivities of amorphous arroys, specific heat of quantum spins, etc.

Fermi-Weighted Time-Dependent Method (FWTDM) [13, 14] is a combination of time-dependent method, random vector representation of trace [15], and Chebyshev polynomial expansion of step function $\theta(E_f - H)$ to extract occupied and unoccupied one-particle states of noninteracting fermi particles in the ground state. This method was successfully applied to optical absorption of silicon and carbon nano-crystallites[14].

Boltzmann-Weighted Time-Dependent Method (BWTDM)[16], calculates linear response functions at finite temperatures by using the Boltzmann-weight operator. The BWTDM has an advantage to FTLM that the computational time does not increase as temperature increases, because BWTDM does not calculate any eigenstates at all. Therefore BWTDM does not need use of symmetries and can easily treat disordered systems.

2. Time-Dependent Schrödinger Equation

With the increasing availability of high performance computers, the development of efficient numerical methods [17, 18] of the time-dependent Schroedinger equation (TDSE) [19]

$$i\frac{d}{dt}|\phi,t\rangle = H|\phi,t\rangle \tag{1}$$

has become an important task in various fields of physics such as scattering theory, quantum chaos, laser-atom/molecule interaction and so on. In this section, we study numerical methods for calculating TDSE with a time-independent Hamiltonian H represented by an $N \times N$ Hermitian matrix and a wave function $|\phi,t\rangle$ represented by N-dimensional complex vector. The atomic units$\hbar = m_e = e = 1$ are used throughout this article. The eigenstates of the Hamiltonian are denoted as

$$H|E_m\rangle = E_m|E_m\rangle, \quad (m = 1, 2, \cdots, N) \tag{2}$$

wile the range of energy spectrum is defined as

$$\Delta E_{grid} = E_{max} - E_{min}, \tag{3}$$

where $E_{max} = \max[\{E_m\}]$ and $E_{min} = \min[\{E_m\}]$ are the largest and smallest eigenvalues of H. Here we define the normalized Hamiltonian by shifting the origin of energy and rescaling the unit of energy.

$$H_{norm} = \frac{H - (E_{max} + E_{min})/2}{(E_{max} - E_{min})/2} \tag{4}$$

which has the eigenvalues in the range of $[-1, +1]$. Hereafter we use H_{norm} in place of H, and write it simply as H by omitting the subscript.

The formal solution of (1) is expressed by the time-evolution operator, i.e., the matrix exponential function [19]

$$|\phi, t + \Delta t\rangle = \exp(-iH\Delta t)\,|\phi, t\rangle \tag{5}$$

Various schemes have been proposed to approximate this exponential function [20, 21, 22]. The simplest scheme called Euler scheme (EU) expands the exponential function to the first order of $H\Delta t$,

$$|\phi, t + \Delta t\rangle = (1 - iH\Delta t)|\phi, t\rangle + O((H\Delta t)^2), \tag{6}$$

and uses (6) repeatedly to obtain $|\phi, t + n\Delta t\langle$. This is an explicit scheme, which does not need matrix inversion, but it is numerically unstable due to the lack of the time reversal symmetry ($t \to -t$). Moreover it is not unitary.

To avoid this instability, the Crank-Nicholson scheme (CN) has been widely used in which the exponential function is approximated by the Caley transform

$$|\phi, t + \Delta t\rangle = \frac{(1 - iH\Delta t/2)}{(1 + iH\Delta t/2)}|\phi, t\rangle + O((H\Delta t)^3), \tag{7}$$

This scheme is unitary, unconditionally stable, and accurate up to $(H\Delta t)^2$. However, this implicit method is prohibitive when the band width of the Hamiltonian is large because the memory and CPU time for solving the linear equation increases as $O(N^2)$. Therefore the development of explicit stable integration methods has been desired.

One of these explicit methods is the Suzuki-Trotter scheme [23, 21].

Another is the symmetrized version of the EU, which is called the second order differencing scheme (MSD2) [22],

$$|\phi, t + \Delta t\rangle - |\phi, t - \Delta t\rangle = -2iH\Delta t|\phi, t\rangle + O((H\Delta t)^3). \tag{8}$$

This scheme is symmetric in time and shown to be conditionally stable [22]. Furthermore, the scheme is accurate up to $(H\Delta t)^2$.

The more accurate and stable method is the Chebyshev scheme (CH) [20]. This method is classified as a global propagator method. It uses very long time steps and sometimes completes the calculation with a single time step.

The CH is very accurate and the error can be reduced up to the machine precision. The only shortcoming of the CH is that the intermediate wave functions are not available, while the stepwise methods, such as the CN and the MSD2, produce the wave function at each time step.

Let us study the stability and error of the various schemes by computing the time-evolution of the eigenstates $|E_m\rangle$. Note that (6-8) contain only unit matrix and Hamiltonian matrix. Therefore, preparing the initial wave function as one of the eigenstates, the wave function remains in the same eigenstates during the time evolution. Then we define the growth factor g as

$$|E_m, t+\Delta t\rangle = g|E_m, t\rangle \tag{9}$$

where g is a complex number. The exact time-evolution should give the growth factor as

$$g = \exp\left(-iE_m\Delta t\right) \tag{10}$$

However, an approximate scheme will give the growth factor of general form

$$g = |g|\exp\left(-iE_m\Delta t + i\varepsilon_{phase}\right). \tag{11}$$

If $|g| \neq 1$ then the scheme becomes numerically unstable and the error in norm grows up exponentially as $|g|^{N_{step}} - 1$, where N_{step} is the number of time steps. If $|g| = 1$ then the norm is conserved and the accumulating error appears only in the phase error ε_{phase}. Note that phase error accumulates after N_{step} time evolutions as

$$\varepsilon_{phase}(t = N_{step}\Delta t) = \varepsilon_{phase}(t = \Delta t)N_{step} \tag{12}$$

which does not growy exponentially but linearly in time. Therefore phase error can be regarded as the error in eigenenergy.

The growth factor equation for EU is obtained by introducing (9) in to (6), $g + i\Delta t - 1 = 0$, where Δt is the dimensionless time step. Then EU is found unconditionally unstable for a finite time step $\Delta t > 0$, $|g| = \sqrt{1+(\Delta t)^2} > 1$.

The growth factor equation of CN becomes $g = \frac{1-i\Delta t/2}{1+i\Delta t/2}$, and CN is found unconditionally stable for any Δt $|g| = 1$

The growth factor equation for MSD2 becomes $g^2 + 2i\Delta t g - 1 = 0$ whose solutions are $g_{\pm} = -i\Delta t \pm \sqrt{1-\Delta t^2}$

If and only if the stability condition $|\Delta t| \leq 1$ is satisfied, the scheme is stable $|g_{\pm}| = 1$.

The phase error under the stability condition can be estimated by the Taylor expansion of the time-evolution operators in (7) and (8),

$$\varepsilon^{CN}_{phase} = \frac{1}{12}(\Delta t)^3 N_{step} \tag{13}$$

$$\varepsilon^{MSD2}_{phase} = -\frac{1}{6}(\Delta t)^3 N_{step} \tag{14}$$

3. Random Vector

A *uniform random vector* is defined by

$$|\Phi\rangle \equiv \sum_{n=1}^{N} |n\rangle \xi_n \tag{15}$$

where $\{|n\rangle\}$ is the basis set used in the computation and ξ_n are a set of random variables generated by a subroutine, which satisfy the statistical relation

$$\langle\langle\, \xi_n \,\rangle\rangle = 0 \tag{16}$$

$$\langle\langle\, \xi_{n'}\xi_n \,\rangle\rangle = 0 \tag{17}$$

$$\langle\langle\, \xi^*_{n'}\xi_n \,\rangle\rangle = \delta_{n'n}. \tag{18}$$

Here $\langle\langle\, \cdot \,\rangle\rangle$ stands for the statistical average.

This random vector may be also expressed by the eigenstates of H,

$$|\Phi\rangle = \sum_{n=1}^{N} |E_n\rangle \zeta_n. \tag{19}$$

Although we do not know the actual value of ζ_n and $|E_n\rangle$, we can derive the statistical relation of ζ_n as follows.

$$\zeta_n = \sum_{l=1}^{N} \langle E_n|l\rangle \xi_l \tag{20}$$

$$\zeta^*_n = \sum_{l=1}^{N} \xi^*_l \langle l|E_n\rangle. \tag{21}$$

Then the statistical relation of ζ_n is derived as

$$\langle\langle\, \zeta_n \,\rangle\rangle = \sum_{l=1}^{N} \langle E_n|l\rangle \langle\langle\, \xi_l \,\rangle\rangle = 0 \tag{22}$$

$$\begin{aligned} \langle\langle\, \zeta^*_{n'}\zeta_n \,\rangle\rangle &= \sum_{l'=1}^{N}\sum_{l=1}^{N} \langle l'|E_{n'}\rangle \langle E_n|l\rangle \langle\langle\, \xi^*_{l'}\xi_l \,\rangle\rangle \\ &= \sum_{l=1}^{N} \langle E_n|l\rangle\langle l|E_{n'}\rangle = \langle E_n|E_{n'}\rangle = \delta_{n'n}. \end{aligned} \tag{23}$$

We can easily notice that the random vector contains all eigenstates of the Hamiltonian with equal probability. Therefore the random vector (15) represents the system at a very high temperature.

The statistical average of $\langle\Phi|X|\Phi\rangle$ gives the trace of X as follows:

$$\begin{aligned} &\langle\langle\, \langle\Phi|X|\Phi\rangle \,\rangle\rangle \\ &= \sum_n X_{nn} + \sum_{n_1,n_2} \langle\langle\, \xi^*_{n_1}\xi_{n_2} - \delta_{n_1 n_2} \,\rangle\rangle\, X_{n_1,n_2} \\ &= \sum_n X_{nn} = \mathrm{tr}\left[X\right]. \end{aligned} \tag{24}$$

The second term in (24) gives the statistical fluctuation when the statistical average is evaluated by using random vectors. The order of the fluctuation is estimated as follows[24]: Assuming all non-zero matrix elements have the value of order 1, the first term in (24) becomes order of N, while the second term becomes order of $\sqrt{N}$ for sparce matrices because the the number of non-zero matrix elements is order N for sparce matrices and the order of fluctuation is proportional to the square root of it. As a result, the relative statistical error will be order of $\sqrt{N}/N = 1/\sqrt{N}$. For example, the statistical error becomes order of 10^{-3} for $N = 10^6$, which can be considered as small. If the statistical fluctuation is not small enough with a set of random vectors, the calculation can be repeated with M sets of random vectors for reducing the fluctuation to order of $1/\sqrt{MN}$. When the spectral density such as DOS and linear response function, the dimension of matrix N in the above estimation should be replaced by N_{eff}, the number of resonaces with in the spectral resolution η, e.g. $N_{eff} = \rho(\omega)\eta$ in case of DOS. Therefore to reach the same accuracy we need more random vectors for higher energy resolution or low temperatures. See Ref. [15] for more sophisticated analysis of this statistical fluctuation.

Next, we try to simulate the Fermi degenerated ground state of a non-interacting many-electron system by a single one-particle wave function, and use this wave function to calculate the linear-response function of the system. We define a *projected random vector* by applying the projection operator $\theta(E_f - H)$ to the random vector,

$$|\Phi_{E_f}\rangle \equiv \theta(E_f - H)|\Phi\rangle. \tag{25}$$

The projected random vector may be also expressed by the basis set $\{|E_n\rangle\}$,

$$|\Phi_{E_f}\rangle = \sum_{E_n \leq E_f} |E_n\rangle\zeta_n \tag{26}$$

where E_f is the Fermi energy. Then the statistical average of $\langle\Phi_{E_f}|X|\Phi_{E_f}\rangle$ gives the sum of contributions from each occupied states

$$\begin{aligned} &\langle\langle\ \langle\Phi_{E_f}|X|\Phi_{E_f}\rangle\ \rangle\rangle \\ &= \sum_{E_n \leq E_f} \langle E_n|X|E_n\rangle = \mathrm{tr}[\theta(E_f - H)X] \end{aligned} \tag{27}$$

Boltzmann-weighted random vectors, $|\Phi_{Boltz}\rangle = e^{-\beta H/2}|\Phi\rangle$, and numerical solution of the time-dependent Schrödinger equation.

4. Random Phase Vector

For numerical evaluation of (24) with K samples of random vectors, the second term of (24) gives the statistical fluctuation,

$$\delta X = \frac{1}{K}\sum_k \sum_{n_1,n_2} \left(\xi_{n_1}^{*(k)}\xi_{n_2}^{(k)} - \delta_{n_1 n_2}\right) X_{n_1 n_2}. \tag{28}$$

In the special case of the Hermitian matrix $X_{n_1 n_2} = X^*_{n_2 n_1}$, it is shown that δX becomes a real number by adding the expression (28) with the subscripts n_1 and n_2 exchanged. The order of the fluctuation for a general matrix X is estimated [25] as

$$\begin{aligned} |\delta X|^2 &= \frac{1}{K^2} \sum_{k,k'} \sum_{n_1 n_2 n_3 n_4} \left(\xi^{*(k)}_{n_1} \xi^{(k)}_{n_2} - \delta_{n_1 n_2} \right) X_{n_1 n_2} \\ &\quad \times \left(\xi^{(k')}_{n_3} \xi^{*(k')}_{n_4} - \delta_{n_3 n_4} \right) X^*_{n_3 n_4} \\ &= \frac{1}{K^2} \sum_{k,k'} \sum_{n_1 n_2 n_3 n_4} \xi^{*(k)}_{n_1} \xi^{(k)}_{n_2} \xi^{(k')}_{n_3} \xi^{*(k')}_{n_4} X_{n_1 n_2} X^*_{n_3 n_4} \\ &+ \frac{1}{K} \sum_{k} \sum_{n_1 n_2 n_3 n_4} \xi^{*(k)}_{n_1} \xi^{(k)}_{n_2} (-\delta_{n_3 n_4}) X_{n_1 n_2} X^*_{n_3 n_4} \\ &+ \frac{1}{K} \sum_{k} \sum_{n_1 n_2 n_3 n_4} (-\delta_{n_1 n_2}) \xi^{(k)}_{n_3} \xi^{*(k)}_{n_4} X_{n_1 n_2} X^*_{n_3 n_4} \\ &+ \sum_{n_1 n_2 n_3 n_4} \delta_{n_1 n_2} \delta_{n_3 n_4} X_{n_1 n_2} X^*_{n_3 n_4}. \end{aligned} \tag{29}$$

By taking the statistical average of (29) and carefully evaluating $\langle\langle \xi^{*(k)}_{n_1} \xi^{(k)}_{n_2} \xi^{(k')}_{n_3} \xi^{*(k')}_{n_4} \rangle\rangle$ by using (16)-(18), we obtain

$$|\delta X|^2 = \frac{1}{K} \left\{ (\langle\langle |\xi_n|^4 \rangle\rangle - 1) \sum_n |X_{nn}|^2 + \sum_{n_1 \neq n_2} |X_{n_1 n_2}|^2 \right\} \tag{30}$$

where the factor $\left(\langle\langle |\xi_n|^4 \rangle\rangle - 1 \right)$ is factored out because we assumed the identical probability distribution for all ξ_n. The factor $1/K$ ensures the behavior $|\delta X| \sim 1/\sqrt{K}$ expected from the central limit theorem. According to the inequality

$$\langle\langle\ |\xi_n|^4\ \rangle\rangle \geq (\langle\langle\ |\xi_n|^2\ \rangle\rangle)^2 = 1, \tag{31}$$

the fluctuation becomes the smallest for a given basis set if and only if $|\xi_n| = 1$ for each n and for each sample of random variables. One of such random vectors is called the *random phase vector* [26, 24, 13] and defined by

$$|\Phi_{random\ phase}\rangle \equiv \sum_{n=1}^{N} |n\rangle e^{i\theta_n} \tag{32}$$

where θ_n are a set of independent uniform random variables defined in $[-\pi, \pi]$. Obviously, the random variables $\xi_n = e^{i\theta_n}$ satisfy the statistical relations (16)-(18). Further, each sample of random phase vectors is automatically normalized without statistical fluctuation,

$$\begin{aligned} &\langle \Phi_{random\ phase} | \Phi_{random\ phase} \rangle \\ &= \sum_{n',n} e^{i(\theta_n - \theta'_n)} \langle n'|n\rangle = \sum_{n=1}^{N} 1 = N. \end{aligned} \tag{33}$$

This result is consistent with the observation in [15] that normalized random vectors give less statistical errors than unnormalized random vectors. It is important to notice that the definition of the random phase vector, $|\xi_n| = 1$, depends on the choice of the basis set. For example, let us examine a unitary transformation of the coefficients, $\xi_i = e^{i\theta_i}$, $(i = 1,2)$, of a random phase vector,

$$\zeta_1 = \frac{1}{\sqrt{2}}\xi_1 + \frac{1}{\sqrt{2}}\xi_2 \tag{34}$$

$$\zeta_2 = \frac{1}{\sqrt{2}}\xi_1 - \frac{1}{\sqrt{2}}\xi_2. \tag{35}$$

The transformed coefficients ζ_n, of course, satisfy the relation (16)-(18), but $|\zeta_n| = 1 \pm \cos(\theta_1 - \theta_2) \neq 1$. Therefore the random phase vector in the original basis set is not a random phase vector in the new basis set. For the random phase vector in a given basis set, the fluctuation (30) reduces to a simple form [24],

$$|\delta X|^2 = \frac{1}{K} \sum_{n_1 \neq n_2} |X_{n_1 n_2}|^2 \tag{36}$$

which becomes zero for diagonal matrices as expected. Note that the fluctuation (36) depends on the choice of the basis set and that it is very important for reducing the fluctuation to choose a basis set that makes off-diagonal matrix elements as small as possible. For a Hermitian matrix X, in theory, we can choose a basis set that diagonalizes X and remove the fluctuation completely.

5. Chebyshev Polynomial

Function of matrix H, f(H), can be expanded by Chebyshev polynomials [11],

$$f(H)|\phi\rangle = \sum_{n=0}^{N_{CH}} a_n T_n(H)|\phi\rangle \tag{37}$$

where $T_n(x)$ are the Chebyshev polynomials of n-th order, and a_n are the expansion coefficients.

In the case of $f(H) = \exp(-iHt)$, $a_0 = J_0(t)$ and $a_n = 2(-i)^n J_n(t), (n > 0)$ [9, 27] where $J_n(x)$ is the Bessel function of the first kind. The order of Chebyshev expansion N_{CH} is taken large enough to obtain good convergence.

In the case of $f(H) = \exp(-\beta H)$, $a_0 = I_0(\beta)$ and $a_n = 2I_n(\beta), (n > 0)$ [9] where $I_n(x)$ is the modified Bessel function of the first kind.

In the case of $f(H) = \theta(E_f - H)$, $a_0 = 1 - \phi_f/\pi$ and $a_n = 2g_m \sin(m\phi_f)/m\pi, (n > 0)$ where g_m is the damping factor [28]

The Chebyshev polynomial of n-th order $T_n(x)$ is calculated by the Chebyshev recursion formula,

$$T_{n+1}(H)|\phi\rangle = 2HT_n(H)|\phi\rangle - T_{n-1}(H)|\phi\rangle. \tag{38}$$

6. Nano-crystallite

In this section we introduce the Projection Method for linear-response function . First we review a simple derivation of the linear-response function by using full many-body state. Second we introduce the *projected random vector*, which simulates the ground states of a non-interacting many-electron system by a single one-particle wave function. Then, combining these two concepts, we derive a fast algorithm for the linear-response function of a non-interacting many-electron system.

The linear-response $\delta B(t)$ to an impulse of perturbation $A\delta(t)$ forced on a quantum system described by the Hamiltonian H is calculated as

$$\delta B(t) \quad = \quad 2\,\mathrm{Im}\langle\phi^{(0)}|e^{+iHt}Be^{-iHt}A|\phi^{(0)}\rangle \tag{39}$$

and the Fourier transformation of $\delta B(t)$ gives the linear response to the perturbation $Ae^{-i\omega t}$

$$\chi_{BA}(\omega+i\eta) = \int_0^T dt\; e^{+i(\omega+i\eta)t}\delta B(t) \tag{40}$$

where the imaginary part of frequency η is introduced to limit the integration time to a finite value $T = -\ln\delta/\eta$ with δ being the relative numerical accuracy of Eq. (40).

By introducing the projected random vector (26) into $|\phi^{(0)}\rangle$ in (39) and then into (40), we obtain our final result, i.e., the linear response of a non-interacting many-electron system,

$$\chi_{BA}(\omega+i\eta) \quad = \quad \left\langle\!\!\left\langle \int_0^T dt\;\; e^{+i(\omega+i\eta)t}\delta B(t) \right\rangle\!\!\right\rangle \tag{41}$$

where $\delta B(t)$ represents the sum of the response from each electron below the Fermi energy, which is defined by

$$\delta B(t) = 2\,\mathrm{Im}\langle\Phi_{E_f}|e^{+iHt}Be^{-iHt}\theta(H-E_f)A|\Phi_{E_f}\rangle. \tag{42}$$

In (42), another projection operator $\theta(H-E_f)$ has been introduced to ensure that the excited states should be higher than the Fermi energy.

For calculating (41) and (42), we start with one realization of the random vector (15), and calculate two wave functions,

$$|\phi^{(0)};t\rangle \quad = \quad e^{-iHt}|\Phi_{E_f}\rangle \tag{43}$$

$$|\delta\phi;t\rangle \quad = \quad e^{-iHt}\theta(H-E_f)A|\Phi_{E_f}\rangle \tag{44}$$

where the time evolution is calculated by the leap frog method [17] and the projection operators are calculated by the Chebyshev polynomial expansion [29, 13]. At each time step, the response (42) and its Fourier transformation (41) are evaluated. Since the leap frog method and the Chebyshev polynomial expansion consist of the matrix-vector operation $H|\phi\rangle$ whose computational efforts are of Order(N) for sparse Hamiltonians, the total computational effort also becomes of Order(N).

Figure 1 shows the calculated absorption coefficient spectra given by $K = 2\omega k/c$ with the complex refractive index defined by $n+ik = \varepsilon^{1/2}$[30].

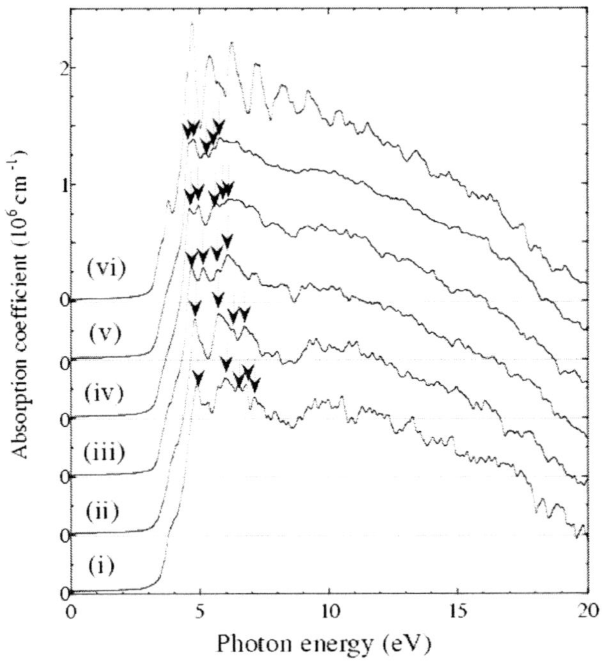

Figure 1. Optical absorption coefficient of hydrogenated Si nanocrystallites (i) $Si_{1050}H_{498}$, (ii) $Si_{1632}H_{666}$, (iii) $Si_{4048}H_{1182}$, (iv) $Si_{8120}H_{1950}$, and (v) $Si_{13464}H_{2730}$ and (vi) bulk Si with 13824 Si atoms. Zero points of the y-axis are shifted as shown by the horizontal lines. Several peak positions are denoted by the arrows[30]

7. Nano-molecular Magnet

In this section, we apply BWTDM to study Electron Spin Resonance spectrum of one-dimensional $s = \frac{1}{2}$ antiferromagnet Cu benzoate $Cu(C_6H_5COO)_2 \cdot 3H_2O$, especially the dynamical crossover between *spinon* excitation and *breather* excitation as a function of temperature. Nonperturbative analytical calculation of this phenomena is a very difficult problem which still open. Even within Sine Gordon theory, the linear response functions have been studied only at high temperature regime (by perturbation) and at zero temperature. We compare our numerical results with preceding experimental and theoretical studies [31, 32, 33, 34, 35, 36, 37, 38, 39, 40, 41].

The effective Hamiltonian of one-dimensional $s = \frac{1}{2}$ antiferromagnet Cu benzoate may be written [38, 39, 40] as

$$H = \sum_{j=1}^{N_{spin}} \left[J\mathbf{s}_j \cdot \mathbf{s}_{j+1} - g\mu_B H_x s_j^x + (-1)^j g\mu_B h s_j^y \right]. \tag{45}$$

The first term is the isotropic Heisenberg antiferromagnet where $\mathbf{s}_j$ is spin operator at the j-th site, $J/k_B = 17.2K = 1.48(meV)$ is the exchange interaction [31], and the sums are taken over N_{spin} spins with periodic boundary conditions. The second is the normal Zeeman term due to the external uniform magnetic field H_x. The third is the staggered Zeeman term due to induced staggered magnetic field $h = 0.065H_x$ originating from the

Dzyaloshinskii-Moriya (DM) interaction and the staggered component of the g tensor[36]. In the following we assume that $g = 2.25(2.29)$ with H_x along $c(c'')$-direction. Unit of magnetic field and frequency become $J/g_c\mu_B = 11.4(T)$, $J/g_{c''}\mu_B = 11.2(T)$, and $J/h = 359(GHz)$, respectively.

According to the linear response theory, absorption of electromagnetic waves of frequency ω and polarization μ is expressed as

$$I^\mu(\omega) = \frac{H_R^2\omega}{2}\chi''_{\mu\mu}(q=0,\omega) \tag{46}$$

where H_R is the amplitude of the electromagnetic waves and $\chi''_{\mu\mu}(q,\omega)$ is the imaginary part of the dynamical magnetic susceptibility,

$$\chi''_{\mu\mu}(q,\omega) = (1-e^{-\beta\omega})\ \mathrm{Im}\lim_{\eta\to+0}\int_0^\infty dt e^{-i(\omega-i\eta)t} g_q^\mu(t) \tag{47}$$

at inverse temperature $\beta = 1/(k_BT)$. The correlation function is defined by

$$g_q^\mu(t) = \mathrm{Tr}\left[e^{-\beta H}M_{-q}^\mu e^{+iHt}M_{+q}^\mu e^{-iHt}\right]/\mathrm{Tr}\left[e^{-\beta H}\right] \tag{48}$$

where the magnetization operator is $M_q^\mu = \sum_{j=1}^{N_{spin}} s_j^\mu \frac{e^{iqj}}{\sqrt{N_{spin}}}$.

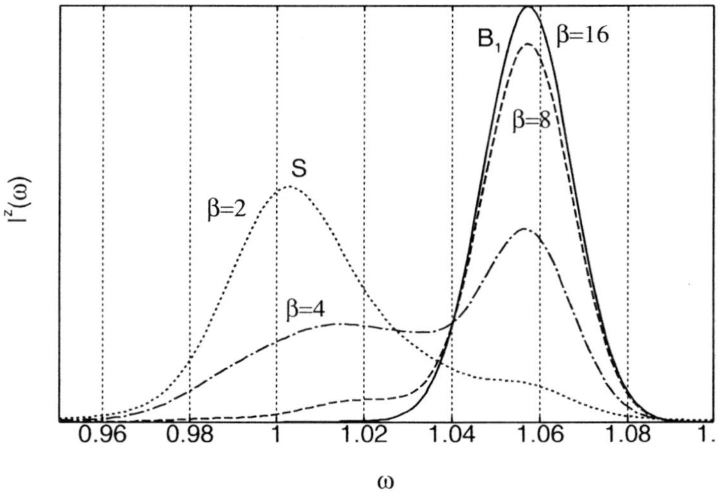

Figure 2. ESR spectra of normal polarization, $I^z(\omega)$, calculated with $\beta = 2 \sim 16$, $H_x = 1.0$, $N_{spin} = 16$, $\eta = 0.01$ and $N_{rand} = 16$. S and B_1 stand for spinon excitation and first breather excitation, respectively

The essence of BWTDM is evaluation of eq. (48)

$$g_q^\mu(t) \quad = \quad \frac{\langle\langle\ \langle\Phi_{Boltz}|M_{-q}^\mu e^{+iHt}|M_{+q}^\mu|e^{-iHt}|\Phi_{Boltz}\rangle\ \rangle\rangle}{\langle\langle\ \langle\Phi_{Boltz}|\Phi_{Boltz}\rangle\ \rangle\rangle}. \tag{49}$$

by using *Boltzmann-weighted random vectors*, $|\Phi_{Boltz}\rangle = e^{-\beta H/2}|\Phi\rangle$, and numerical solution of the time-dependent Schrödinger equation.

In real calculation, we first generate N complex random numbers ξ_n and construct $|\Phi\rangle$ according to eq. (15). The $|\Phi_{Boltz}\rangle$ can be computed with numerical stability by applying eq. (37) repeatedly to $|\Phi\rangle$. Then $\langle\Phi_{Boltz}|\Phi_{Boltz}\rangle$ gives a sample of the denominator of eq. (49). Here, we introduce another vector $|\Phi_{M^{\mu}_{+q}}\rangle = M^{\mu}_{+q}|\Phi_{Boltz}\rangle$ and calculate the time evolution of $|\Phi_{Boltz}\rangle$ and $|\Phi_{M^{\mu}_{+q}}\rangle$ by eq.(8). At each time $t = n\Delta t$, the matrix element $\langle\Phi_{M^{\mu}_{+q}};t|M^{\mu}_{+q}|\Phi_{Boltz};t\rangle$ gives a sample of numerator of eq. (49). After calculating the denominator and numerator for N_{rand} random vectors, the ratio of their averages gives $g^{\mu}_{q}(t)$ of eq. (49). Then $\chi''_{\mu\mu}(q,\omega)$ with a finite frequency resolution η is calculated by using Gaussian filter,

$$\chi''_{\mu\mu}(q,\omega) = (1 - e^{-\beta\omega})\,\mathrm{Im}\int_0^{T_{max}} dt e^{-i\omega t} g^{\mu}_{q}(t)\left[Ae^{-\eta^2 t^2/2}\right] \tag{50}$$

where T_{max} is related to η by $T_{max} \sim \eta^{-1}$. In order to avoid finite size effect of spinon excitations, T_{max} is chosen so that $T_{max} < N_{spin}/v_{spin}$ where $v_{spin} \sim 1$ is the spin wave velocity [40]. This limits the finest frequency resolution. Note that this is not the limit due to the algorithm but due to the finite size model. Much finer resolution can be used for breather modes, which are spatially localized.

Figure 2 shows the ESR spectra of normal polarization (Faraday configuration) $I^z(\omega)$ calculated with various temperatures. At high temperatures ($\beta \leq 5$), S is outstanding in the spectrum, and its peak shifts to higher frequency and becomes broadened as temperature decreases while B_1 becomes narrower. At low temperatures ($\beta \geq 5$), B_1 prevails while its peak frequency is almost constant and its width becomes much smaller than the numerical resolution η. This crossover behavior between spinon excitation and first breather excitation has become computable by the invention of BWTDM, and the result is consistent with both experimental [32] and field-theoretical results [36]. Several other weak peaks appear in our results as well as in experimental results, which are supposed to be higher-order breathers and transitions between excited states. However, we are not going to analyze them here.

In summary, we developed an efficient and stable algorithm for linear response functions at finite temperatures, and studied the ESR spectra of $s = \frac{1}{2}$ antiferromagnet Cu benzoate for a wide range of temperature and magnetic field. We reproduced experimental results of the spinon-breather crossover as a function of temperature. Temperature dependence of the width and shift of the peaks are also calculated, which are consistent with experiments and analytical theories. The calculated frequency of B_1 as a function of H_x agrees well with both experimental and field-theoretical results at low fields, and reproduces the deviation of experimental results from the Sine Gordon theory at high fields, where the low energy assumption of the Sine Gordon theory may be broken and the choice of the compactification radius R becomes ambiguous[37]. The advantage of BWTDM is being applicable to finite temperatures, strong magnetic field and high frequency while its weak point is finite size effects ($N_{spin} \leq 20$). The computational cost of BWTDM is moderate. Calculation of one curve in Fig. 1, for example, requires approximately 30 minutes with 8 CPU's of Fujitsu VPP5000 vector-parallel computer.

Acknowledgments

The results presented here were computed by using supercomputers at RIKEN and NIG.

References

[1] M. Suzuki, S. Miyashita, and A. Kuroda, *Prog. Theor. Phys.*, (1977), 58, 1377.

[2] W. Linden, *Phys. Rep.*, (1992), 220, 53.

[3] R.N. Silver, J.E. Gubernatis, and D.S. Sivia, *Phys. Rev. Lett.*, (1990), 65, 496.

[4] C. Lanczos, *J. Res. Nat. Bur. Stand.*, (1950), 45, 255.

[5] E. Dagotto, *Rev. Mod. Phys.*, (1994), 66, 763.

[6] J. Jaklic and P. Prelovsek, *Phys. Rev. B*, (1994), 49, 5065.

[7] R.N. Silver and H. Roeder, *Int. J. Mod. Phys. C*, (1994), 5, 735.

[8] L.W. Wang and A. Zunger, Solving schroedinger's equation around a desired energy: Application to silicon quantum dots, *J. Chem. Phys.*, (1994), 100(3), 2394.

[9] R. Kosloff and H. Tal-Ezer, *Chem. Phys. Lett.*, (1986), 127, 223.

[10] W. Kunishima, T. Tokihiro, and H. Tanaka, Error estimation of recursive orthogonal polynomial expansion method for large hamiltonian system, *Comp. Phys. Comm.*, (2002), 148, 171-181.

[11] W.H. Press, S.A. Teukolsky, W.T. Vetterling, and B.P. Flannery, *Numerical Recipes in Fortran 77*, Cambridge University Press, 1994.

[12] M.L. Williams and H.J. Maris, *Phys. Rev. B*, (1985), 31, 4508.

[13] T. Iitaka, S. Nomura, H. Hirayama, X Zhao, Y. Aoyagi, and T. Sugano, O(n) algorithm for linear response functions of non-interacting electrons, *Phys. Rev. E*, (1997), 56, 1222 (*cond-mat/9703224*).

[14] S. Nomura, T. Iitaka, X. Zhao, Y Aoyagi, and T Sugano, Linear scaling calculation for optical-absorption spectra of large hydrogenated silicon nanocrystallites, *Phys. Rev. B*, (1997), 56(8), R4348.

[15] A. Hams and H. de Raedt, Fast algorithm for finding the eigenvalue distribution of very large matrices, *Phys. Rev. E*, (2000), 62, 4365.

[16] T. Iitaka and T. Ebisuzaki, Algorithm for linear response functions at finite temperatures: Application to esr spectrum of $s = \frac{1}{2}$ antiferromagnet cu benzoate, *Phys. Rev. Lett.*, (2003), 90, 047203 (*cond-mat/0212485*).

[17] T. Iitaka, Solving the time-dependent schrodinger equation numerically, *Phys. Rev. E*, (1994), 49, 4684.

[18] Toshiaki Iitaka, Parity Physics Course, *Introduction to Quantumdynamics*, Maruzen, Tokyo, 1994.

[19] J.J. Sakurai, *Modern Quantum Mechanics*, Addison Wesley, Massachusetts, 1994, (*revised version*).

[20] C. Leforestier, R.H. Bisseling, C. Cerjan, M.D. Feit, R. Friesner, A. Guldberg, A. Hammerich, G. Jolicard, W. Karrlein, H.-D. Meyer, N. Lipkin, O. Roncero, and R. Kosloff, *J. Comp. Phys.*, (1991), 94, 59.

[21] H. de Raedt, *Comp. Phys. Rep.*, (1987), 7, 1.

[22] A. Askar and A.S. Cakmak, *J. Chem. Phys.*, (1978), 68(6), 2794.

[23] M. Suzuki, *Phys. Lett. A*, (1990), 146, 319.

[24] T. Iitaka, Computing the real-time green's functions of large hamiltonian matrices, *High Performance Computing in RIKEN*, (1996) (*physics/9802021*).

[25] T. Iitaka and T. Ebisuzaki, Random phase vector for calculating the trace of a large matrix, *Phys. Rev. E*, (2004), 69, 057701.

[26] R. Alben, M. Blume, H. Krakauer, and L. Schwartz, Exact results for a three-dimensional alloy with site diagonal disorder: comparison with the coherent potential approximation, *Phys. Rev. B*, (1975), 12(10), 4090.

[27] A. Vijay and H. Metiu, *J. Chem. Phys.*, (2002), 116, 60.

[28] R.N. Silver, H. Roeder, A.F. Voter, and J.D. Kress, *J. Comp. Phys.*, (1996), 124, 115.

[29] O.F. Sankey, D.A. Drabold, and A. Gibson, Projected random vectors and the recursion method in the elecronic-structure problem, *Phys. Rev. B*, (1994), 50, 1376.

[30] S. Nomura, T. Iitaka, X. Zhao, , T. Sugano, and Y. Aoyagi, Quantum size effect in model nanocrystalline/amorphous mixed-phase silicon structures, *Phys. Rev. B*, (1999), 59(15), 10309.

[31] K. Oshima, K. Okuda, and M. Date, *J. Phys. Soc. Jpn.*, (1978), 44, 757.

[32] T. Asano, et al., *Phys. Rev. Lett.*, (2000), 84, 5880.

[33] A. Ogasahara, S. Miyashita, *Prog. Theor. Phys. Suppl*, (2002), 145, 286.

[34] M. Oshikawa and I. Affleck, *Phys. Rev. Lett.*, (1999), 82, 5136.

[35] I. Affleck and M. Oshikawa, *Phys. Rev. B*, (1999), 60, 1038.

[36] M. Oshikawa and I. Affleck, *Phys. Rev. B*, (2002), 65, 134410.

[37] M. Oshikawa, *Prog. Theor. Phys. Suppl.*, (2002), 145, 243.

[38] M. Oshikawa and I. Affleck, *Phys. Rev. Lett.*, (1997), 79, 2883.

[39] F.H.L. Essler and A.M. Tsvelik, *Phys. Rev. B*, (1998), 57, 10592.

[40] F.H.L. Essler, *Phys. Rev. B*, (1999), 59, 14376.

[41] J. Lou, et al., *Phys. Rev. B*, (2002), 65, 064420.

In: Molecular Simulation Studies in Material ... ISBN 1-59454-607-X
Editor: Kholmirzo Kholmurodov, pp. 67-83

Chapter 7

First Principles Molecular Dynamics Calculation for Planetary and Earth Sciences: $MgSiO_3$ Post-Perovskite and CH_4 Hydrate

*Toshiaka Iitaka**
Computational Astrophysics Laboratory
RIKEN (The Institute of Physical and Chemical Research)
2-1 Hirosawa, Wako, Saitama 351-0198, Japan

Abstract

One of the difficulties in studying the internal structures of the earth and other planets comes from the fact that there are no direct methods for observing inside the planets. For decades the technique of laboratory experiment at high pressure and high temperature equivalent to those inside planets has been rapidly developing and providing new information that is unavailable by observations only. The first principle molecular dynamics calculation is recently attracting astrophysicists' attention as the third method, which is complementary to indirect observation and laboratory experiment. Here we review our latest study of first principle molecular dynamics focused on the two newly discovered crystal structures, $MgSiO_3$ post-perovskite and methane (CH_4) hydrate phase III.

1. Introduction

The evolution of the Earth and the life on her surface has been greatly influenced by the physical and chemical properties of the materials that compose the planet. For example [1, 2], about 250 million years ago, geologists believe, a huge plume of hot mantle was formed deep inside the Earth, near the boundary of the mantle and core, where a mass of strange mineral, so-called *D" layer* exists. The plume comes up, reaches to the crust and touches the *methane hydrate* buried under the bottom of the ocean. Then the large scale

*E-mail address: tiitaka@riken.jp

melting of methane hydrate releases a huge amount of methane gas, which is twenty times more effective in global warming than carbon dioxide is. The increase of methane gas in the atmosphere is supposed to have caused a drastic change of global climate and the greatest biological mortality event, in which 95 percent of biological species was extinguished.

In the following sections, we review our recent study of first principle molecular dynamics calculation focused on the two newly discovered crystal structures, $MgSiO_3$ post-perovskite (the most promising candidate of mineral composing the D" layer) [3] and methane hydrate phase III (a high pressure phase of methane hydrate that might solve the mystery of Jupiter's satellite Titan) [4, 5].

2. $MgSiO_3$ Post-perovskite

$MgSiO_3$ perovskite has been assumed to be the dominant component of the Earth's lower mantle, although this phase alone cannot explain the discontinuity in seismic velocities observed 200-300 km above the core-mantle boundary (the D" discontinuity) or the polarization anisotropy observed in the lowermost mantle [6]. Experimental and theoretical studies that have attempted to attribute these phenomena to a phase transition in the perovskite phase have tended to simply confirm the stability of the perovskite phase [7, 8, 9, 10, 11]. However, recent in situ X-ray diffraction measurements have revealed [12] a transition to a 'post-perovskite' phase above 125 GPa and 2,500 K-conditions close to those at the D" discontinuity. Here we show the results of first-principles calculations of the structure, stability and elasticity of both phases at zero temperature. We find that the post-perovskite phase becomes the stable phase above 98 GPa, and may be responsible for the observed seismic discontinuity and anisotropy in the lowermost mantle. Although our ground-state calculations of the unit cell do not include the effects of temperature and minor elements, they do provide a consistent explanation for a number of properties of the D" layer.

Figure 1. The unit cell structures of $MgSiO_3$. a) Perovskite; b) post-perovskite. The spheres represent Mg, and the octahedra represent Si with sixfold oxgen coordination

The first-principles calculations were performed with the initial models of the perovskite and post-perovskite phases, which are the orthorhombic unit cells containing four molecules with the space group *Pbnm* and *Cmcm*, respectively (Fig. 1). Then the lattice constants and the positions of the atoms were optimized within the given symmetry to minimize the enthalpy (see *Methods*). This procedure was repeated by changing the external pressure from 80 GPa to 130 GPa in steps of 10 GPa. It was found that the post-perovskite phase becomes the stable phase above 98 GPa at T=0 K by drawing a graph of the enthalpy difference $\Delta H = H(PP) - H(Pv)$ as a function of pressure (Fig. 2). (Here we use PP to indicate a property of the post-perovskite phase, and Pv a property of the perovskite phase.)

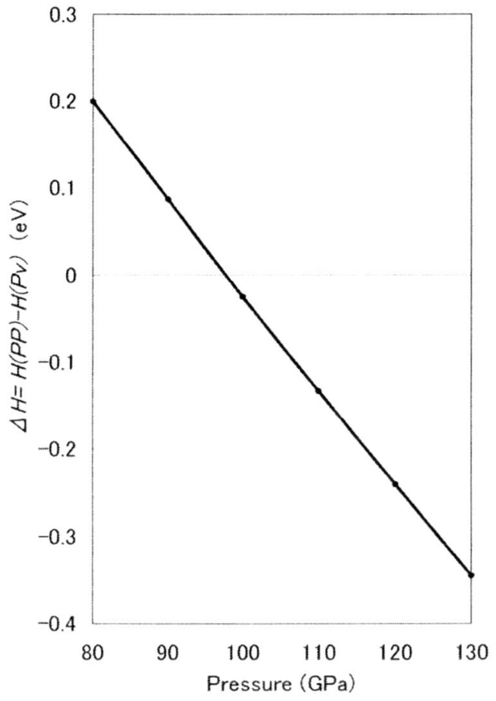

Figure 2. The enthalpy difference between the perovskite phase and post-perovskite phase as a function of pressure. The post-perovskite (PP) phase is favoured over the perovskite (Pv) phase at pressures above 98 GPa

Table 1 lists the parameters of the optimized structures of the post-perovskite and perovskite phases at 120 GPa, together with the experimental data [12]. The coordination numbers of Mg and Si in the post-perovskite phase are same as those in the perovskite phase, namely, eight for Mg and six for Si. The principal difference between the two structures is the linkage of SiO_6 octahedra. In the post-perovskite phase, they share edges, thereby forming chains, and the chains share the octahedral corners thereby forming flat sheets [12], whereas in the perovskite phase, the octahedra share corners, thereby forming a three-dimensional network. The average Si-O distance of the post-perovskite phase (1.665 Å) is longer than that of the perovskite phase (1.650 Å); on the other hand, the average Mg-O distance of the post-perovskite phase (1.959 Å) is shorter than that of the perovskite phase (1.981 Å). The reduction of the Mg-O polyhedra and both the edge sharing and deformation of the SiO_6 octahedra cause a volume reduction of the post-perovskite phase from that of the perovskite phase. The small space for cations suggests that Mg, which is a very small cation, is much more comfortable in post-perovskite phase than other cations. Our first-principles calculations show that the post-perovskite phase is about 1.4 % denser than the perovskite phase throughout the pressure range, which is in agreement with the experimental value7 of 1.0-1.2 % at 120 GPa and 300 K.

The post-perovskite phase of $MgSiO_3$ (Fig. 1b) is isostructural with $UFeS_3$, $UFeSe_3$, $ThMnTe_3$, $ThMgTe_3$, $UMnSe_3$ and $ThMnSe_3$ (refs 8–10). The crystal habit of $ThMnTe_3$ and $ThMgTe_3$ was reported [14] to be platy (and needle-like). Although the crystallographic direction was not mentioned in the literature, the surface of the plate should be the (010) surface, which is parallel to the sheet structure, and the direction of the needle should be the [100] direction, which is the direction of the edge-shared octahedral chain. There is an important difference between the post-perovskite phase of $MgSiO_3$ and those of $Th^{(IV)}Mn^{(II)}Te_3^{(II)}$ and $Th^{(IV)}Mg^{(II)}Te_3^{(II)}$: the eight-coordinated Mg–O bonds are weaker than $Th^{(IV)}$–$Te^{(II)}$ and the octahedral Si–O bonds are stronger than $Mn^{(II)}$–$Te^{(II)}$ and $Mg^{(II)}$–$Te^{(II)}$. These bonding features imply that the inter-chain bonds of the $MgSiO_3$ post-perovskite phase are much stronger than those of the $ThMnTe_3$ and $ThMgTe_3$ post-perovskite phase, and that the $MgSiO_3$ post-perovskite phase most probably has the platy crystal habit with (010) facets. The system of uranium transition-metal chalcogenides UMS_3 and $UMSe_3$ (where M is a transition metal) also shows an interesting relation to the perovskite–post-perovskite phase transition, because the structure of these compounds changes when M changes, instead of when the pressure changes: UMS_3 (M=V, Cr, Co, Ni, Ru, Rh) and $UMSe_3$ (M=V, Cr, Co, Ni) take the *Pbnm* structure, and $UMnSe_3$, $UFeSe_3$ and $UFeS_3$ take the *Cmcm* structure. The mechanism behind this structure change is still unknown [15].

Table 2 shows the nine elastic stiffness constants c_{ij} evaluated by the first-principles calculations. The crystal of the post-perovskite phase is much more compressible along the b-axis than along the a- or c-axis as $c_{22} << c_{11}$, c_{33}, which could be a significant source of its seismic anisotropy. The propagation direction ($\mathbf{n}$) dependence of the seismic wave velocities in the perovskite and post-perovskite phase at the phase boundary ($\sim$100 GPa) were evaluated by solving the Christoffel equationdet $\left|c_{ijkl}n_j n_l - \rho V^2 \delta_{ik}\right| = 0$ for the single crystal (Fig. 3). The wave velocities vary significantly with the propagation direction and the polarization, suggesting that both the phases exhibit strong anisotropy in both compressional (P) and shear (S) waves. The particular softness of the post-perovskite phase along the b-axis is reflected in the velocity minima when the polarization is along the b-axis, that is, when the propagation direction is along [010] for P waves or along [100] and [001] for S waves. A measure of anisotropy may be defined [16, 17] by A=$(v_{max} - v_{min})/< v > \times 100\%$, where v_{max} and v_{min} are the maximum and minimum velocities, respectively, and ¡v¿ is the velocity for an isotropic aggregate of crystal calculated from the c_{ij} by the Voigt-Reuss-Hill averaging scheme [6]. Though both the phases show strong anisotropy, the post-perovskite phase is found more anisotropic than the perovskite phase: the azimuth anisotropy of P and S waves is A_P=16% and A_S=31% for the post-perovskite phase, whereas A_P=11% and A_S=16% for the perovskite phase.

The D” discontinuity is observed [18] in many regions around the world about 200–300 km above the core–mantle boundary with an increase of up to 3.0% in both S- and P-wave velocities. On the basis of global seismic tomography, Sidorin [19] proposed a

Table 1. Crystal data of post-perovskite and perovskite phases.

		Post-perovskite phase		Perovskite phase	
		Experimental [12] (121 GPa, 300 K)	Calculated (120 GPa, 0 K)	Calculated (120 GPa,0 K)	
Crystal system		Orthorhombic	Orthorhombic	Orthorhombic	
Space group		*Cmcm* (no. 63)	*Cmcm*	*Pbnm* (no. 62)	
Cell param.					
a (Å)		2.456	2.455	4.289	
b (Å)		8.042	8.051	4.557	
c (Å)		6.093	6.099	6.264	
Z		4	4	4	
V (Å^3)		120.39	120.55	122.43	
Atomic coord.					
Mg (4c)	x	0.0	0.0	0.477	
	y	0.253	0.253	0.925	(4c)
	z	0.25	0.25	0.25	
Si (4a)	x	0.0	0.0	0.0	
	y	0.0	0.0	0.0	(4a)
	z	0.0	0.0	0.0	
O1 (4c)	x	0.0	0.0	0.884	
	y	0.923	0.928	0.033	(4c)
	z	0.25	0.25	0.25	
O2 (8f)	x	0.0	0.0	0.817	
	y	0.631	0.636	0.307	(8d)
	z	0.436	0.442	0.557	
Interatomic dist. (Å)					
Si–O			1.632(×2), 0.681(×4)	1.645(×2) 1.649(×2) 1.657(×2)	
Average			[1.665]	[1.650]	
O–O edges			2.296, 2.305, 2.327, 2.337, 2.380, 2.456	2.313, 2.321, 2.330, 2.346, 2.347, 2.349, 2.415, 2.464	
Mg–O			1.867(×2), 1.943(×4) 2.084(×2)	1.815, 1.848(×2), 1.885, 2.042(×2) 2.185(×2)	
Average			[1.959]	[1.981]	
Shortest Si–Si			2.455	3.129	
Si–Mg			2.547	2.494	
Mg–Mg			2.455	2.997	

model of a possibly ubiquitous seismic discontinuity atop the D" layer that is caused by a solid–solid phase transition with a Clapeyron slope of 6 MPa K^{-1}. The increase of seismic velocities atop the D" layer is attributed to the increase of bulk (K) and shear (G) moduli at the transition, which overcomes the increase of the density. One of the weak points of this model was that no relevant phase transition had been observed in the major constituents of the lower mantle, $MgSiO_3$ perovskite and magnesiowüstite. Therefore the discovery of the post-perovskite phase transition above 125 GPa and 2,500 K by Murakami *et al.* [12] provides a strong ground for the model. The phase boundary drawn with the transition pressures at T=0 K (from the first-principles calculation) and at T=2,500 K (from the high-pressure experiments) has a Clapeyron slope of 10 MPa K^{-1}, which is close to Sidorin's value.

In the D" layer significant polarization anisotropy has been reported [20, 21], especially under the circum-Pacific regions, where the horizontally polarized S-wave velocity (v_{SH}) is 1–3% faster than the vertically polarized S-wave velocity (v_{SV}). The post-perovskite phase transition can explain why the polarization anisotropy is observed only in the D" layer but not in the overlying mantle [20]. In the possible horizontal shear flow in the D" layer, crystals of the post-perovskite phase are probably preferentially oriented with the most compressible *b*-axis in the vertical direction. This seems very likely because the crystal of the post-perovskite phase has the sheet stacking structure along the *b*-axis (Fig. 1b). Therefore we modelled the phase boundary by using an isotropic aggregate of perovskite phase and a transversely isotropic aggregate of post-perovskite phase with the *b*-axis as the symmetry axis. Then we calculated the seismic wave velocities at the phase boundary by using the Voigt-Reuss-Hill averaging scheme for the perovskite phase, and the scheme of refs 17 and 18 for the velocities of the horizontally propagating P wave ($v_{PH}^{(PP)}$) and the horizontally and vertically polarized S-waves ($v_{SH}^{(PP)}$, $v_{SV}^{(PP)}$) in the post-perovskite phase. The discontinuities in P and S waves are estimated as 4–7% or $v_{PH}^{(PP)}/< v_P^{(Pv)} >=1.04$, $v_{SH}^{(PP)}/< v_S^{(Pv)} >=1.07$, and $v_{SV}^{(PP)}/< v_S^{(Pv)} >=1.03$. The polarization anisotropy of S waves in the post-perovskite phase is estimated as 4% or $v_{SH}^{(PP)}/v_{SV}^{(PP)}=1.04$. This result may be regarded as consistent with the observations [18, 19] if we consider the effect of imperfect crystal alignments. With isotropic aggregates of the post-perovskite phase, very small discontinuities, $< v_P^{(PP)} >/< v_P^{(Pv)} >=0.999$ and $< v_S^{(PP)} >/< v_S^{(Pv)} >=1.01$, and no anisotropy are expected. Our model is in contrast to the model with the preferentially oriented perovskite phase where the most compressible axis becomes horizontal, and therefore the wrong polarization anisotropy, $v_{SV} > v_{SH}$, occurs [23, 24].

Moreover, anti-correlation between bulk-sound velocity and S-wave velocity has been observed in the very deep portions of the mantle [25]. The present results show slower bulk-sound velocity and faster S-wave velocity for the post-perovskite phase than for the perovskite phase at equivalent pressure, suggesting that such negative correlation may be accounted for by the heterogeneity in the mineral proportion of the perovskite phase and the post-perovskite phase. It is, however, noted that the present calculations were made at T=0 K and for the Mg end-member composition. The effects of temperature [26, 27] and the solution of minor elements were neglected here. The experiments on natural pyrolitic mantle composition showed that the $MgSiO_3$-rich post-perovskite contains much less iron than the perovskite [28], in accordance with our 'space for cations' argument.

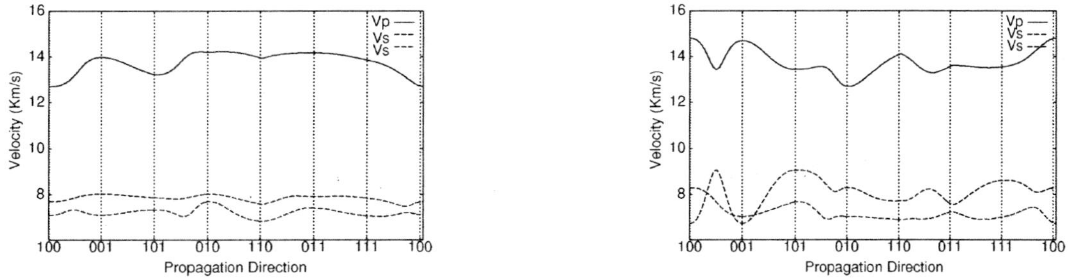

Figure 3. The variation of compressional (v_P) and shear (v_S) wave velocities as a function of propagation direction. **a**, Perovskite phase at 100 GPa; **b**, post-perovskite phase at 100 GPa.

Table 2. The elastic-stiffness constants c_{ij} of post-perovskite and perovskite phases.

	Post-perovskite		Perovskite		Perovskite [23]	
	100 GPa	120 GPa	100 GPa	120 GPa	100 GPa	120 GPa
cc_{11}	11175	11270	8852	9909	9926	9996
cc_{22}	8862	9937	11068	11160	11160	11263
cc_{33}	11157	11264	11033	11128	11056	11156
cc_{12}	3366	4425	4461	5520	4460	5530
cc_{13}	2289	3329	3368	4410	3380	4415
cc_{23}	4437	4493	3391	4434	4406	4449
cc_{44}	2264	2291	3340	3363	3360	3388
cc_{55}	2242	2264	2265	2278	2265	2279
cc_{66}	3368	4412	3312	3339	3330	3360
K	595	660	596	654	623	682
G	306	332	296	314	314	337

Methods

Our first-principles calculations were performed with CASTEP 4.2 codes based on the plane wave basis set, the Vanderbilt-type ultrasoft pseudopotentials [29, 30, 31] for electron–ion interaction, and the local density approximation (LDA) for exchange-correlation interaction. Previous studies [16, 17, 23, 31] show that such a method reproduces very well the structures and properties of silicates under high pressure. The Brillouin zones are sampled with 6×4×4 Monkhorst-Pack k-points [32] for the perovskite phase and 8 ×2×4 k-points for the post-perovskite phase by using the maximal symmetry of the model. Increasing the number of k-points to 6×6×4 and 10×4×4, respectively, changes the total energy only 10^{-5} eV per atom. The cut-off energy is set to 800 eV; increasing the cut-off energy to 1,600 eV changes the enthalpy difference of the two phases by only 10^{-3} eV per atom. During the structural optimization, the enthalpy $H=E+PV$ is minimized by varying cell shape and atomic positions under the restriction of the given symmetry. In the geometrical optimization, the total stress tensor [33] is reduced to the order of 0.01 GPa by using the finite basis-set corrections [34]. The nine elastic stiffness constants c_{ij} are evaluated by computing the stress tensor σ_i generated by forcing strain ε_j to the optimized unit cell with four different magnitudes, ε=-0.02, -0.01, 0.01 and 0.02, and by fitting them to the third-order polynomial of ε to remove the nonlinear contributions [16, 17, 35].

3. Methane Hydrate

Methane hydrate (MH), known as *Burning Ice*, is a special class of ice that contains methane molecules in cages or networks of hydrogen bonded water molecules. Low pressure phase of methane hydrate (MH-I) forms sI structure of cages [36]. MH-I, abundant in the deep ocean, has been attracting attention of the industry as a key material of new energy resource, whose amount is estimated twice as much as the total fossil fuel reserve [36].

MH is also known as an important material for understanding the mystery of the atmosphere of Titan, the largest satellite of Saturn. The conventional theory [37] could not explain abundant methane gas in Titan's atmosphere because MH-I inside Titan was assumed to decompose into ice and methane around 1 or 2 GPa, and escape to the atmosphere to be photodecomposed in early stage of Titan's history. To understand this mystery is one of the goals of the Cassini-Huygens spacecraft, which started its journey in 1997 and will arrive at Saturn system in 2004 [38].

On the earth, in 2001, Loveday *et al.* [39, 40] discovered new phases of MH by X-ray and neutron diffraction experiments under high pressure: MH-I transforms to MH-II (sH cage structure) at 1 GPa, and then to MH-III phase (orthorhombic filled ice structure) at 2 GPa, which survives at least up to 10 GPa. Other researchers reported similar high pressure phases [42, 41]. Recently Hirai *et al.* [43] reported that MH-III survives up to 42 GPa at room temperature. Shimizu *et al.* [44] have measured the site- and pressure-dependence of CH- and OH-vibration frequencies in these phases up to 5.2 GPa. Discovery of these high pressure phases allows us a new explanation of abundant methane gas in Titan's atmosphere: methane gas may be reserved in thick layers of MH-III under Titan's surface and gradually emit to the atmosphere from the reservoir [39].

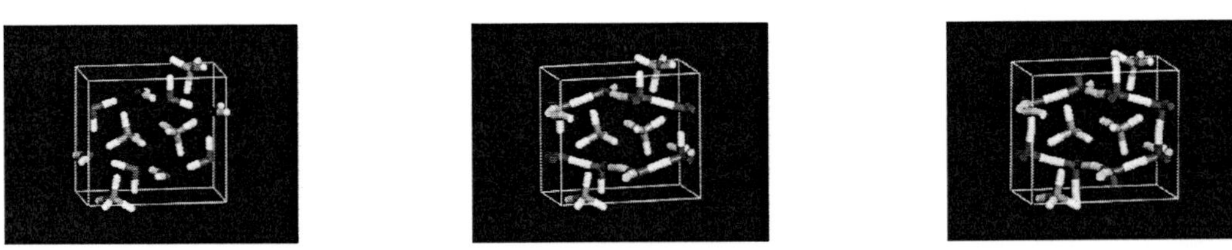

Figure 4. Crystal Structure of MH-III (a) 40 GPa, (b) 60 GPa, (c) 80 GPa

In this article, we focus on the features of MH-III as a promising material for investigating *centering* or symmetrization of hydrogen bonds between water molecules. In spite of a long quest to understand this phenomenon [45, 46, 47, 48, 49, 51, 52, 53, 50, 54, 55], a unified theoretical explanation of various experimental results appeared only recently [47], partly because of difficulties in measurements under very high pressures. Advantage of using MH-III for studying the centering is that it is expected to occur at much lower pressure (70 GPa) than that of pure ice (ice VII-ice X transition around 100 GPa), which may make the difficult high-pressure experiments easier. In the following, we calculate the crystal structure and vibrational spectra of MH-III by using the density functional theory, so that we can predict and analyze experimental results.

We modeled the crystal structure of MH-III at 3 GPa by using the *Pmcn* symmetry, the lattice parameters and the position of atoms determined by the diffraction experiment [40]. We chose the orientation of methane molecules carefully so that the molecules become close packed. The model consists of four methane and eight water molecules in the unit cell, of which one methane and two water molecules are symmetrically inequivalent. This structure was found stable after full geometrical optimization, in which the enthalpy $H = E + PV$ was minimized by varying the lattice vectors and the positions of atoms without any constraints such as crystal symmetry. The structures at higher pressures (Figs. 4) were calculated in a similar manner.

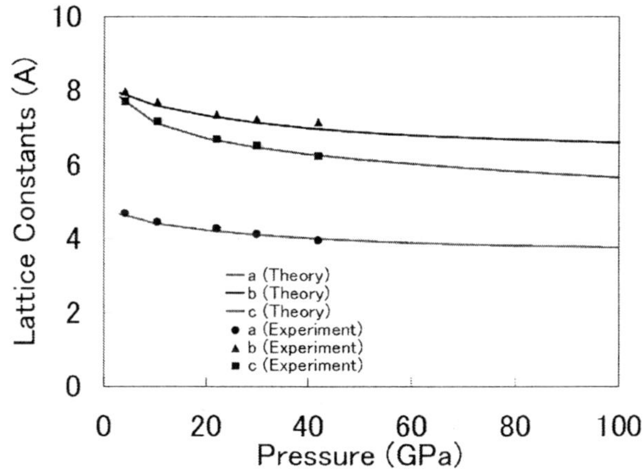

Figure 5. Lattice parameters: The solid lines indicate the present density functional theory calculation. The symbols indicate the experimental results of Hirai *et al.* [43]

The details of the electronic structure calculation [56] are as follows: the valence wave functions are expanded in a plane wave basis set truncated at a kinetic energy of 1520 eV. The electron-ion interactions are described by the Vanderbilt-type ultrasoft pseudopotentials [57]. The effects of exchange-correlation interaction are treated within the generalized gradient approximation of Perdew *et al.* (GGA-PBE) [58]. The Brillouin zones are sampled with $4 \times 2 \times 2$ Monkhorst-Pack k-points [59] by using time-reversal symmetry only. Convergence was checked with respect to cut-off energy and number of k-points. Increasing cut-off energy to 1520 eV changed the total energy 5 meV per atom, and increasing the number of k-points to 48 changed only 0.01 meV per atom. In the geometrical optimization, the total stress tensor is reduced to the order of 0.01 GPa by using the finite basis-set corrections [60].

Figure 5 shows the pressure dependence of the calculated lattice parameters at the zero temperature. It agrees well with the experimental data at room temperature by Hirai et al. [43]. The cell is the most soft along the c-axis. This feature is also evident from the Figures 1. The compression of the cell along the c-axis is mainly caused by the flattening of the two graphite-like wrinkled sheets normal to c-axis. As a result, the hydrogen bond network at very high pressures becomes sp_2-like structure in contrast to sp_3 structure of ice Ih.

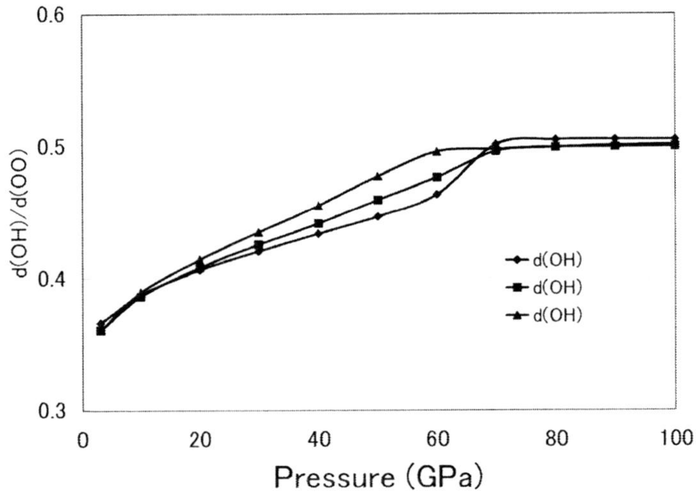

Figure 6. Hydrogen position: The ratio $d(OH)/d(OO)$ calculated by density functional theory is shown as a function of pressure.

In pure ice, centering of the hydrogen bond $O-H\cdots O$ occurs as the oxygen-oxygen distance $d(OO)$ of two hydrogen-bonded water molecules decreases with pressure increase [45, 46]. At large $d(OO)$ the proton occupies one of two potential minima along the $O-H\cdots O$ bond. As $d(OO)$ becomes smaller the barrier between the two minima becomes lower. When $d(OO)$ becomes smaller than some critical distance, the proton occupies the single minimum at the midpoint between the two oxygen atoms (centering). In the following, we neglect the effect of finite temperature and quantum nature of hydrogen atom unless otherwise stated.

Indeed Figures 4 show that the centering occurs around this pressure. Figure 6 shows the ratio $d(OH)/d(OO)$ as a function of pressure. The ratio starts from 0.35 (water molecule) at 3 GPa and reaches to 0.5 (centering) around 70 GPa.

Now, let us look into the thermal effects. Benoit *et al.* [47] recently proposed *three-stage scenario* for explaining the hydrogen bond centering [48] of pure ice with increase of pressure at room temperature: under low pressure ice stays in *molecular state* where H_2O remain water molecule, then under medium pressures hydrogen atoms start to jump between two potential minima (*ionized state*), finally under high pressures hydrogen atoms sit at the midpoint between the oxygen atoms (*centering state*). From this figure we can read that the ionization and centering in MH-III are expected around 40 GPa and 70 GPa, respectively. The probability distributions of $d(OH)/d(OO)$ for a hydrogen bond of MH-III at room temperature and with pressure 3 GPa, 40 GPa, and 80 GPa are calculated with Car-Parrinello molecular dynamics [61] (Fig. 7), which clearly shows that *three stage scenario* is also valid in MH-III but with much lower pressure than pure ice.

Figure 8 shows the pressure dependence of normal mode frequencies of MH-III at $T=0$ K calculated with the density functional linear response theory [62]. The modes between 3000 cm^{-1} and 3500 cm^{-1} are the CH vibrations, whose frequencies monotonically increase with pressure. The modes between 2000 cm^{-1} and 3000 cm^{-1} are OH-stretching

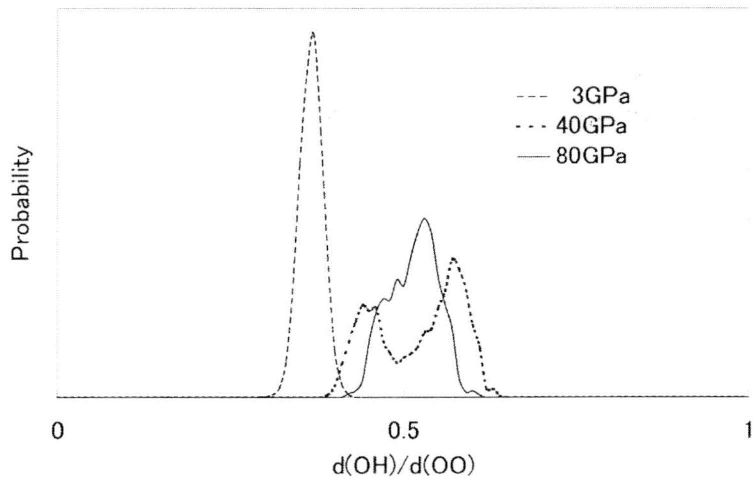

Figure 7. Distribution of hydrogen atom in a hydrogen bond of MH-III at 300K.

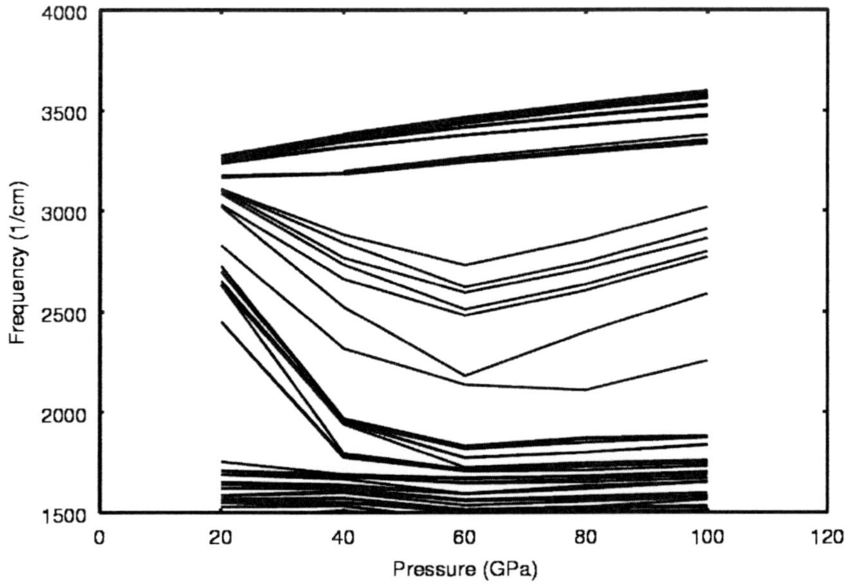

Figure 8. Normal mode frequencies calculated with the density functional linear response theory as a function of pressure.

modes. Their frequencies decrease monotonically up to 70 GPa, where centering occurs, and then start to increase. This tendency is qualitatively in accordance with that of pure ice [49].

We calculated vibrational spectra of MH-III at 3 GPa (molecular state) for T=300 K and T=30 K as a Fourier transform of velocity-velocity correlation function obtained from Car-Parrinello molecular dynamics [61] and compared with the normal mode frequencies at T=0 K calculated with the density functional linear response theory [62]. The peak around 3100 cm^{-1} characteristic to OH-stretching vibration of hydrogen bonded water molecules is prominent at 30 K but disappears at 300K. This fact is consistent with the experimental observation that the Raman peak around 3100 cm^{-1} disappeared in MH-III at room temperature [44], and can be interpreted as the result of strongly anharmonic potential. The disappearance of the OH-stretching Raman peak is also observed in simulations of ionized states. In vibrational spectra at 80 GPa (centering state), the OH-stretching modes of centered hydrogen bonds are located between 2000 cm^{-1} and 3000 cm^{-1}. At low temperature the peaks agrees well with the normal modes, while the peaks are blue-shifted at room temperature probably due to weak anharmonicity. The disappearance of the peak of the OH-stretching mode in molecular and ionized states at room temperature and the survival of the peak in centered hydrogen bond in MH-III are analogous to those in ice VII and ice X [52].

We calculated also the phonon dispersion of MH-III at 80 GPa with the density functional linear response theory [62]. All frequencies at all wave numbers have positive frequencies indicating the mechanical stability of this structure at zero temperature. However, further experimental and theoretical studies are necessary to prove the thermal stability at room temperature.

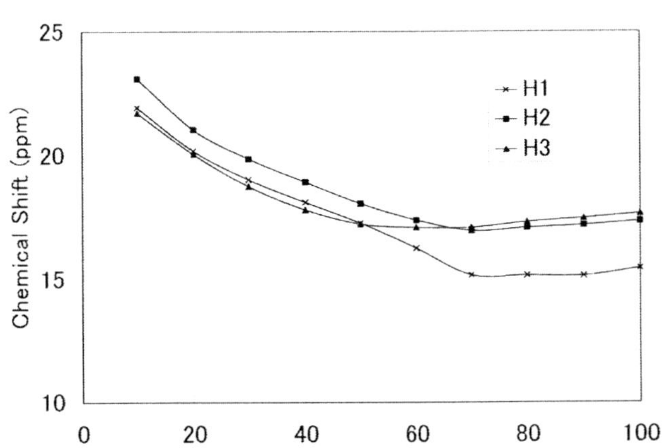

Figure 9. ^{1}H-NMR chemical shift of water molecules in MH-III

Figure 9 shows the ^{1}H-NMR chemical shift of water molecules in MH-III, which is calculated with the density functional linear response theory [63]. The result is consistent with intuitive interpretation of hydrogen bond symmetrization: At low pressure, proton is bound to an oxygen atom and screened by the electrons of the oxygen atom. As pressure increase, symmetrization happens and proton moves to the midpoint between the oxygen atoms where electron density is small. Further increase of pressure makes $d(OO)$ smaller and the electron density at the midpoint becomes gradually larger. Using this picture, symmetrization might be observed by ^{1}H-NMR when NMR measurement at such high pressure become feasible.

4. Summary

We reviewed our latest application of first principle molecular dynamics to planetary and earth sciences with two examples of newly discovered crystal structures, $MgSiO_3$ post-perovskite and methane (CH_4) hydrate phase III.

First, the seismic velocities propagating in the post-perovskite, a new mineral phase discovered recently in the laboratory experiment, were calculated by using the first principles electronic structure calculation. The results can explain many strange features of the Earth's lowermost mantle (D" layer) consistently. This gives a solid ground for the model that the post-perovskite is the main component of the D" layer.

Second, we studied the structural and spectral properties of high pressure phase of methane hydrate (MH-III) with the density functional theory, and showed that ionization and centering of hydrogen bonds in MH-III may be observed around 40 GPa and 70 GPa, respectively, which are much lower pressure than those of pure water ice (55 GPa and 100 GPa). Therefore MH-III may provide useful information about ionization and centering of hydrogen bonds between water molecules as the second example after pure ice. From the view point of planetary science, studying physical and chemical properties of clathrate hydrate such as MH will become more and more important to interpret and to understand the information of the outer solar systems sent from space-missions such as Cassini-Huygens spacecraft arriving at Saturn system.

These examples show that the first principle molecular dynamics calculation is a promising as the third method for studying planetary science, which is complementary to indirect observation and laboratory experiment.

Acknowledgments

The author would like to thank all of the collaborators. The results presented here were computed by using supercomputers at RIKEN and NIG.

References

[1] Svensen, H., Planke, S., Malthe-sorenssen, A., Jamtveit, B., Myklebust, R., Eidem, T.R., Rey, S.S., *Nature*, Release of methane from a volcanic basin as a mechanism for initial Eocene global warming, (2004), 429, 542-545.

[2] Lo, C.H., Chung, S.L., Lee, T.Y., Wu, G.Y., *Earth and Planetary Science Letters*, Age of the Emeishan flood magmatism and relations to Permian-Triassic boundary events, (2002), 198, 449-458.

[3] Iitaka, T., Hirose, K., Kawamura, K., and Murakami, M., *Nature*, The elasticity of the $MgSiO_3$ post-perovskite phase in the Earth's lowermost mantle, (2004), 430, 442-445.

[4] Iitaka, T., and Ebisuzaki, T., *Phys. Rev. B*, Methane hydrate under high pressure, (2003), 68, 172105.

[5] Iitaka, T., and Ebisuzaki, T., *J. Phys.: Conden. Matt.*, Filled ice structure of gas hydrates; a density functional study, (2004), 16, S1171.

[6] Poirier, J. P., *Introduction to the Physics of the Earth's Interior*, Cambridge Univ. Press, Cambridge, 2000.

[7] Fiquet, G., Dewaele, A., Andrault, D., Kunz, M. and Le Bihan, T., *Geophys. Res. Lett.*, Thermoelastic properties and crystal structure of $MgSiO_3$ perovskite at lower mantle pressure and temperature conditions, (2000), 27, 21-24.

[8] Andrault, D., *J. Geophys. Res.*, Evaluation of (Mg, Fe) partitioning between silicate perovskite and magnesiowustite up to 120 GPa and 2300 K, (2001), 106, 2079-2087.

[9] Stixrude, L. and Cohen, R. E., *Nature*, Stability of orthorhombic $MgSiO_3$ perovskite in the Earth's lower mantle, (1993), 364, 613-616.

[10] Wentzcovitch, R. M., Ross, N. L. and Price, G. D., *Phys. Earth Planet. Inter.*, Ab initio study of $MgSiO_3$ and CaSiO3 perovskites at lower-mantle pressures, (1995), 90, 101-112.

[11] Warren, M. C., Ackland, G. J., Karki, B. B. and Clark, S. J., *Mineral. Mag.*, Phase transitions in silicate perovskites from first principles, (1998), 62, 585-598.

[12] Murakami, M., Hirose, K., Kawamura, K., Sata, N. and Ohishi, Y., *Science*, Post-perovskite phase transition in $MgSiO_3$, (2004), 304, 855-858.

[13] Noel, H. and Padiou, J., *Acta Crystallogr.*, Structure crystalline de FeUS3, (1976), B 32, 1593-1595.

[14] Narducci, A. A. and Ibers, J. A., *Inorg. Chem.*, The related compounds $MThTe_3$(M=Mn, Mg) and $ACuThSe_3$ (A=K, Cs): Syntheses and characterization, (2000), 39, 688-691.

[15] Ijjaali, I., Mitchell, K., Huang, F. Q. and Ibers, J. A., *J. Solid State Chem.*, Syntheses and characterization of the actinide manganese selenides $ThMnSe_3$ and $UMnSe_3$, (2004), 177, 257-261.

[16] Karki, B. B. et al., *Am. Mineral.*, Structure and elasticity of MgO at high pressure, (1997), 82, 51-60.

[17] Karki, B. B. et al., *Am. Mineral.*, Elastic properties of orthorhombic $MgSiO_3$ perovskite at lower mantle pressures, (1997), 82, 635-638.

[18] Wysession, M. E. et al., *IN: The Core-Mantle Boundary Region*, eds. Gurnis, M., Wysession, M. E., Knittle, E. and Buffett, B. A., 273-297, American Geophysical Union, Washington DC, 1998.

[19] Sidorin, I., Gurnis, M. and Helmberger, D. V., *Science*, Evidence for a ubiquitous seismic discontinuity at the base of the mantle, 286, 1326-1331 (1999).

[20] Lay, T., Williams, Q., Garnero, E. J., Kellogg, L. and Wysession, M. E., *In: The Core-Mantle Boundary Region*, Eds. Gurnis, M., Wysession, M. E., Knittle, E. and Buffett, B. A., 299-318, American Geophysical Union, Washington DC, 1998.

[21] Panning, M. and Romanowicz, B., *Science*, Inferences on flow at the base of Earth's mantle based on seismic anisotropy, (2004), 303, 351-353.

[22] Montagner, J. P. and Nataf, H. C., *J. Geophys. Res.*, A simple method for inverting the azimuthal anisotropy of surface waves, (1986), 91, 511-520.

[23] Wentzcovitch, R. M., Karki, B. B., Karato, S. and Da Silva, C. R. S., *Earth Planet. Sci. Lett.*, High pressure elastic anisotropy of $MgSiO_3$ perovskite and geophysical implications, (1998), 164, 371-378.

[24] Karato, S., Zhang, S. and Wenk, H. R., *Science*, Superplasticity in Earth's lower mantle: evidence from seismic anisotropy and rock physics, (1995), 270, 458-461.

[25] Masters. G. and Laske G., *In: Earth's Deep Interior: Mineral Physics and Tomography from the Atomic to the Global Scale*, Eds. Karato, S., Forte, A., Liebermann, R., Masters, G. and Stixrude, L., 63-87, American Geophysical Union, Washington DC, 2000.

[26] Oganov, A. R., Brodholt, J. P. and Price, G. D., *Nature*, The elastic constants of $MgSiO_3$ perovskite at pressures and temperatures of the Earth's mantle, (2001), 411, 934-937.

[27] Wentzcovitch, R. M., Karki, B. B., Cococcioni, M. and de Gironcoli, S., *Phys. Rev. Lett.*, Thermoelastic properties of $MgSiO_3$-perovskite: insights on the nature of the Earth's lower mantle, (2004), 92, 018501.

[28] Murakami, M., *Thesis*, Phase transition of lower mantle mineral and its geophysical implications, Tokyo Institute of Technology, 2004.

[29] Vanderbilt, D., *Phys. Rev. B*, Soft self-consistent pseudopotentials in a generalized eigenvalue formalism, (1990), 41, 7892-7895.

[30] Civalleri. B. and Harrison, N.M., *Mol. Simulat.*, New ultrasoft pseudopotentials for the study of silicates, (2002), 28, 213-237.

[31] Brodholt, J. P., Organov, A. R. and Price, G. D., *Phil. Trans. R. Soc. Lond.*, Computational mineral physics and the physical properties of perovskite, (2002), A 360, 2507-2520.

[32] Monkhorst, H. J. and Pack, J. D., *Phys. Rev. B*, Special points for Brillouin-zone integrations, (1976), 13, 5188-5192.

[33] Nielsen, O. H. and Martin, R. M., *Phys. Rev. Lett.*, First principles calculation of stress, (1983), 50, 697-700.

[34] Francis, G. P. and Payne, M. C., *J. Phys. Condens. Matter*, Finite basis set corrections to total energy pseudopotential calculations, (1990), 2, 4395-4404.

[35] Iitaka, T. Ebisuzaki, T., *Phys. Rev. B*, First-principles calculation of elastic properties of solid argon at high pressures, (2002), 65, 012103.

[36] E. D. Sloan, *Clathrate Hydrates of Natural Gases*, Marcel Dekker, 1998.

[37] J. Lunine, D. Stevenson, *Astrophys. J. Suppl.*, (1985), 58, 493.

[38] N. Jet Propulsion Laboratory, http://saturn.jpl.nasa.gov/.

[39] J. Loveday, R. Nelmes, M. Guthrie, S. Belmonte, D. Allan, D. Klug, J. Tse, Y. Handa, *Nature*, (2001a), 410, 661.

[40] J. Loveday, R. Nelmes, M. Guthrie, D. Klug, J. Tse, *Phys. Rev. Lett.*, (2001b), 87, 215501.

[41] I. Chou, A. Sharma, R. Burruss, R. Hemley, A. Goncharov, L. Stern, S. Kirby, *J. Phys. Chem. A*, (2001), 105, 4664.

[42] H. Hirai, Y. Uchihara, H. Fujihisa, M. Sakashita, E. Katoh, K. Aoki, K. Nagashima, Y. Yamamoto, T. Yagi, *J.Chem.Phys.*, (2001), 115, 7066.

[43] H. Hirai, T. Tanaka, T. Kawamura, Y. Yamamoto, T. Yagi, Retention of filled ice structure of methane hydrate up to 42 GPa, *Phys. Rev. B*, (2003), 68, 172102.

[44] H. Shimizu, T. Kumazaki, T. Kume, S. Sasaki, *J. Phys. Chem. B*, (2002), 106, 30.

[45] V. F. Petrenko, R. W. Whitworth, Physics of Ice, Oxford University Press, 1999.

[46] G. A. Jeffrey, An Introduction to Hydrogen Bonding, Oxford University Press, 1997.

[47] M. Benoit, A. Romero, D. Marx, *Phys. Rev. Lett.*, (2002), 89, 145501.

[48] P. Loubeyre, R. LeToullec, E. Wolanin, M. Han, D. Hausermann, *Nature*, (1999), 397, 503.

[49] A. Goncharov, V. Struzhkin, H. Mao, R. Hemley, *Phys. Rev. Lett.*, (1999), 83, 1998.

[50] M. Benoit, D. Marx, M. Parrinello, *Nature*, (1998), 392, 258.

[51] V. Struzhkin, A. Goncharov, R. Hemley, H. Mao, *Phys. Rev. Lett.*, (1997), 78, 4446.

[52] K. Aoki, H. Yamawaki, M. Sakashita, H. Fujihisa, *Phys. Rev. B*, (1996), 54, 15673.

[53] K. Hirsh, W. Holzapfel, *J. Chem. Phys.*, (1986), 84, 2771.

[54] M. Bernasconi, P. Silvestrelli, M. Parrinello, *Phys. Rev. Lett.*, (1998), 81, 1235.

[55] A. Putrino, M. Parrinello, *Phys. Rev. Lett.*, (2002), 88, 176401.

[56] M. C. Payne, M. P. Teter, D. C. Allan, T. A. Arias, J. D. Joannopoulos, *Rev. Mod. Phys.*, (1992), 64, 1045.

[57] D. Vanderbilt, *Phys. Rev. B*, (1990), 41, 7892.

[58] J. P. Perdew, K. Burke, M. Ernzerhof, *Phys. Rev. Lett.*, (1996), 77, 3865.

[59] H. J. Monkhorst, J. D. Pack, *Phys. Rev. B*, (1976), 13, 5188.

[60] G. P. Francis, M. C. Payne, *J. Phys. Condensed Matter*, (1990), 2, 4395.

[61] D. Marx, J. Hutter, Modern Methods and Algorithms in Quantum Chemistry, 1, NIC Series, Forschungzentrum, 2000.

[62] X. Gonze, *Phys. Rev. B*, (1997), 55, 10337.

[63] D. Sebastiani M. Parrinello, *J. Phys. Chem. A*, (2001), 105, 1951.

In: Molecular Simulation Studies in Material ...
Editor: Kholmirzo Kholmurodov, pp. 85-113

ISBN 1-59454-607-X

Chapter 8

MOLECULAR DYNAMICS SIMULATION AND EXPERIMENTAL STUDIES ON THE VISUAL PIGMENT RHODOPSIN: MULTIPLE CONFORMATIONAL STATES AND STRUCTURAL CHANGES

Kh. T. Kholmurodov*
Department of Radiation and Radiobiological Research
Joint Institute for Nuclear Research, 141980, Dubna, Moscow region, Russia
T. B. Feldman†
Biochemical Physics Institute
Russian Academy of Sciences, 17997 Kosygin st. 4, Moscow, Russia
M. A. Ostrovsky
Department of Radiation and Radiobiological Research
Joint Institute for Nuclear Research, 141980, Dubna, Moscow region, Russia

Abstract

Molecular Dynamics (MD) simulations and spectral studies were performed on visual pigment rhodopsin to investigate its dark-adapted and transition states, intermediates and conformation properties. Based on the MD simulations with a supercomputer and special-purposes MDGRAPE-2 machine we investigated for rhodopsin molecule the structural conformational changes and dynamical behaviors. We performed 3 ns MD-calculations on the rhodopsin and presented the structure analysis data for its dark-adapted state. We have compared the dynamical properties of the rhodopsin with regard to the following considerations: (1) Rhodopsin without chromophore retinal; (2) Rhodopsin with chromophore (11-*cis*) retinal. In the first approach for the retinal chromophore all kind interactions inside the rhodopsin pocket as well as with the water solvent were neglected. The RMSD (root-mean-square deviation) simulation data for the helices I-VII of rhodopsin (without retinal) were performed. From the analysis

*E-mail address: mirzo@jinr.ru
†E-mail address: feldman@sky.chph.ras.ru

of the RMSD results we observed a relative movement of rhodopsin helices, which possesses different behavior between the helices. The structure images of rhodopsin were constructed and the representative configurations were compared for the 3 ns state. The results showed the largest displacements from the reference structure for the helices III and V. In the second consideration we have fulfilled the RMSD and structural analysis for the rhodopsin (with 11- *cis* retinal), generated the pictures of the atomic-scale processes for the binding pocket, surrounding the chromophore retinal, and compared the helical deviations for the beta-ionone ring and Schiff base linkage regions of the protein. We have observed that the inclusion of a small retinal chain into the whole rhodopsin dynamics (+ solvent) does not influence the RMSD behavior of the total rhodopsin structure. Within the 3-ns calculations the RMSD of the rhodopsin (with 11-*cis* retinal) possesses the same average value and behavior as in the absence of the retinal chromophore. However, it should be stressed that with regard to selected rhodopsin helices the inclusion of the retinal chromophore changes all RMSD values and causes essential helical distortions.

The most remarkable point of our observations is that the rhodopsin helical distortions in the dark state are accompanied with the transformation of the retinal chromophore, viz. with the rotation of the beta-ionone ring inside the protein binding pocket. From our simulation data we observe that within a short time [0.3-0.4] ns the beta-ionone ring changes its orientation in relation to the polyene chain (chromophore retinal of 11-*cis* conformation). The beta-ionone ring at later stages of the dynamics takes a more planar position with regard to the helix III of the rhodopsin. Within this time interval we observe that the amino acid residue Phe-261 changes its orientation relatively to the retinal chromophore (in beta-ionone region) as well. The positions of the two key amino acid residues Phe-212 and Ala-269 change with the beta-ionone ring rotation. Starting from this interval [0.3-0.4] ns the residue Ala-269 approaches more close to the retinal site (beta-ionone part) than the residue Phe-212.

The low-temperature absorption spectroscopy technique has been used to study the primary stages of rhodopsin photolysis. The digitonin extract of bovine rhodopsin was irradiated at -155 °C by blue light (436 nm). The following dark spectral changes have been recorded in course of a slow temperature rise. The simultaneous appearance of more than one spectral product has been revealed throughout bato- to lumirhodopsin transition in the temperature interval from -155 °C to -145 °C. An intermediate product which is transformed along with batorhodopsin to the next product and known as a blue-shifted intermediate [1] was observed. The structural transformation properties of rhodopsin were discussed in due course of the stimulations and experimental observations. It is proposed that the appearance of more than one intermediate at each stage of rhodopsin photolysis reflects the existence of multiple conformation states of the retinal chromophore in rhodopsin molecule in the dark-adapted state. As a summary of our observations we suggest that the conformation dynamics of the opsin helices determines the most probable chromophore (11-*cis* retinal) configuration, so that a favorable retinal conformation has to be adjusted by the protein surroundings.

Introduction

Visual pigment rhodopsin is the only light-sensitive protein in the disk membrane of the rod outer segment. This molecule belongs to a seven transmembrane helical protein receptor, known as G-protein-coupled receptors family. In Fig. 1 the ribbon structures of rhodopsin with retinal chromophore (twin structure model) are shown.

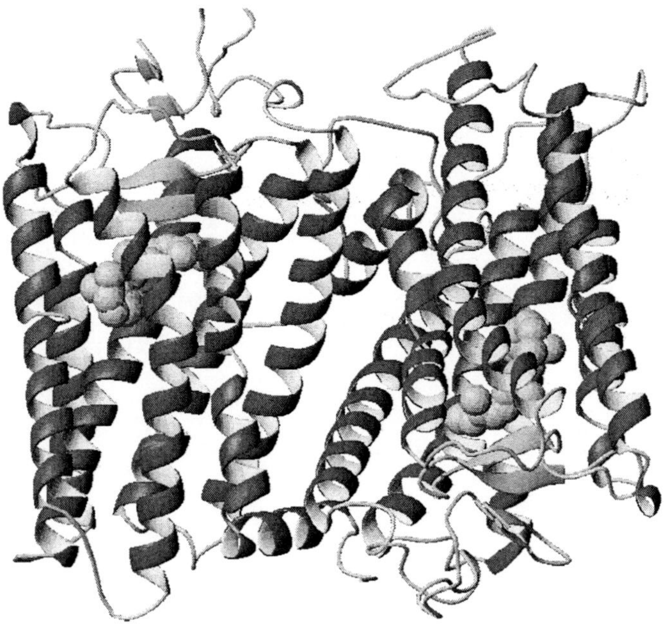

Figure 1. Ribbon structure of the rhodopsin with the retinal chromophores: the retinal polyene chains are drawn by balls. The twin structure model (left: chain A; right: chain B) was constructed using a refined crystal structure of the rhodopsin (PDB entry 1HZX; Teller et. al., 2001; Palczewski et al., 2000).

Rhodopsin consists of 11-*cis* retinal chromophore and protein opsin. 11-*cis* retinal covalently binds to the protein through a protonated Schiff base linkage, namely to the ε-amino group of Lys-296 in the center of the helix VII (see Fig. 2, and [2] for the review).

Light isomerizes the 11-*cis* retinal to its all-*trans* form, that is followed by series of protein conformational changes. Rhodopsin is transformed into its activated state, which interacts with G protein and initiates an enzyme cascade leading to the visual transduction. Finally, the all-*trans* retinal moves from the opsin chromophore site. This process is known as rhodopsin photolysis.

Spectroscopic methods have been extensively applied to the study the reactions following rhodopsin light excitation (see Fig. 3 for a schematic representation). One way to characterize the intermediates is to trap them at low temperatures [2, 3]. This approach led to the classical bleaching scheme (max values in nanometers):

$$rhodopsin\ (Rh)\ (498) \longrightarrow bathorhodopsin\ (Batho)\ (542) \longrightarrow lumirhodopsin\ (Lumi)\ (497)$$
$$\longrightarrow metarhodopsin\ I\ (Meta\ I)\ (480) \longrightarrow metarhodopsin\ II\ (Meta\ II)\ (380)$$

Time-resolved spectroscopy of rhodopsin with femtosecond resolution had allowed identifying the new first product of rhodopsin photolysis, namely photorhodopsin (PhotoRh) (max 585) [4, 5, 6]. Other time-resolved measurements, of up to 1 msec duration, reveal a new intermediate, called blue-shifted intermediate (BSI) [1]. The following reaction scheme was proposed and today is under consideration:

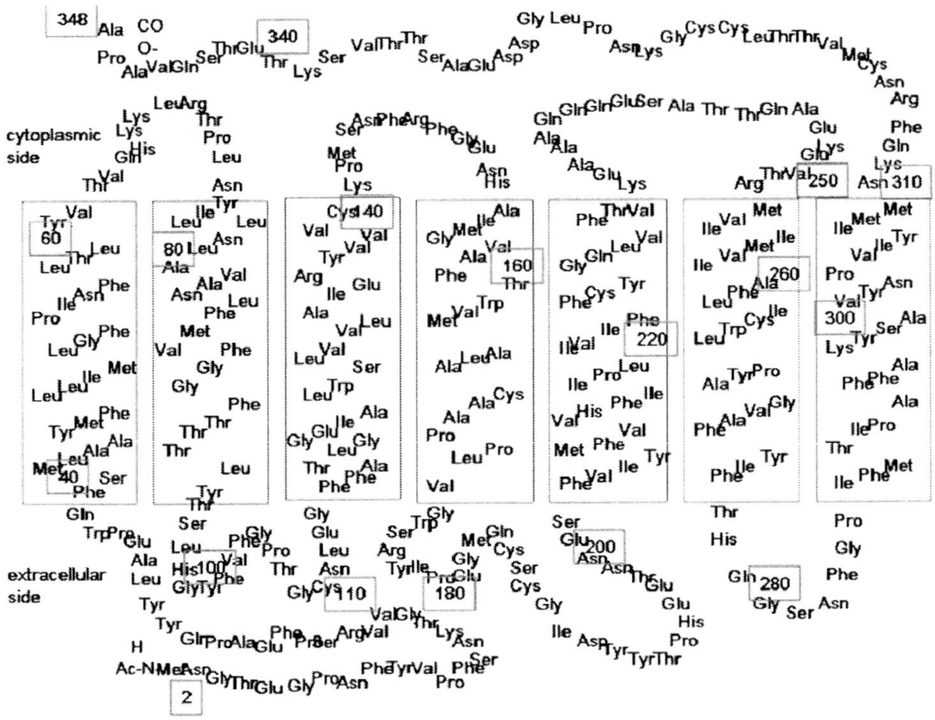

Figure 2. Primary structure diagram with the positions of the helices I-VII (from left to right) of rhodopsin protein is shown (Ovchinnikov et al., 1982; Teller et al., 2001; Palczewski et al., 2000).

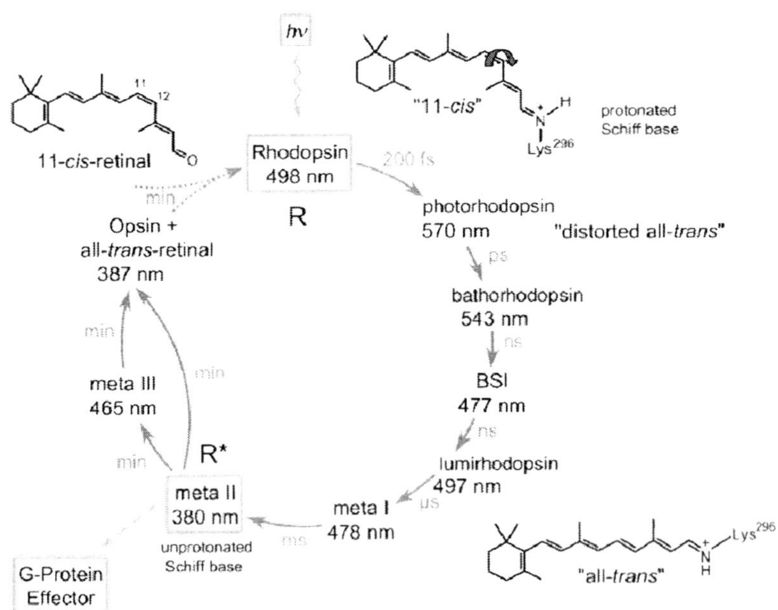

Figure 3. Photocycle of rhodopsin with spectroscopically detected intermediates. (From: O.P. Ernst and F.J. Bartl. Active state of Rhodopsin. Chem. Bio. Chem. 2002. vol .3, pp. 968-974)

$$Rh\ (498) \longrightarrow PhotoRh\ (585) \longrightarrow Batho\ (542) \longleftrightarrow BSI\ (477) \longrightarrow Lumi\ (497)$$
$$\longrightarrow Meta\ I\ (480) \longrightarrow Meta\ II\ (380)$$

Recently resolved crystal structure of rhodopsin provides detailed structural information on the interactions between the retinal chromophore and its surrounding residues [7]. These data have been becoming basis for the Molecular Dynamics Simulations to calculate the rhodopsin dark-adapted state as well as the possible pathways and structures of intermediates after rhodopsin photo excitation.

The correlation of rhodopsin conformational dynamics and activation process of rhodopsin at later stages are the most challenging targets for computer molecular simulations. The relative movement of the rhodopsin helices, for example, could be a part of the rhodopsin light-activated photocycle [43, 34]. The experiments revealed that a relative movement of the rhodopsin helices (for example, a movement of helix VI relative to other helices) plays a key role in the rhodopsin activation [59, 60]. This evidently motivates the extremely time consuming atomic modeling of rhodopsin. It has been considered that large-scale motion of rhodopsin protein may be a common dynamical motion for other alpha-helical G-protein receptors [34, 61].

In today's molecular modeling, rhodopsin conformation and its photo-induced isomerization were mostly studied by the conventional and quantum-mechanical molecular dynamics (MD and *ab initio* QM/MM) methods [34, 40, 41, 42, 43]. Although the *ab initio* QM provides a basic insight into the mechanisms of the rhodopsin conformation changes, the actual reaction paths can deviate due to thermal fluctuations, reaction excess energies, etc. Through the MD simulation, on the other hand, the characteristic of the incoherent kinetic and dynamics can be captured directly [43]. Consequently, conventional MD method represents itself as a more adequate approach for providing a detailed picture of the atomic-scale processes of rhodopsin conformation. Many of the recent MD simulations were aimed to describe rhodopsin isomerization process and its conformational relaxation steps [40, 42, 43]. In [43], for example, MD simulations were performed on the rhodopsin protein embedded in a lipid bilayer and water environment. The retinal chain was isomerized around C_{11}=C_{12} double bond, changing its conformation from 11-*cis* to all-*trans*, in accordance to the experimental observations [44, 45, 46, 47, 48, 49]. It is worth noting, however, that the nature of non-reaction transformation, molecular mechanisms or conformational changes of rhodopsin protein is still unknown. The dynamics of rhodopsin conformation, its photoisomerization processes or transition states may essentially be influenced by the environmental conditions (opsin + solvent) [38, 41].

In Figure 4 the schematic diagrams of the retinal binding pocket for the rhodopsin protein are presented. It is worth noting that the analysis of the electron density data on the 11-*cis*-retinal chromophore and trans-membrane helices [30, 36] provides some preliminarily picture of the intramolecular interactions in the rhodopsin protein. The structural conformations of rhodopsin essentially determine its function as a light-activated receptor.

The retinal binding site or salt bridges in rhodopsin are evidently correlate with the activation mechanism of the protein. The rhodopsin activation is associated, for example, with a shift of the relative orientation of the helices III and VI [34, 43]. The Schiff base group and Glu-113 binding stabilizes the connection between the helices III and VII, the

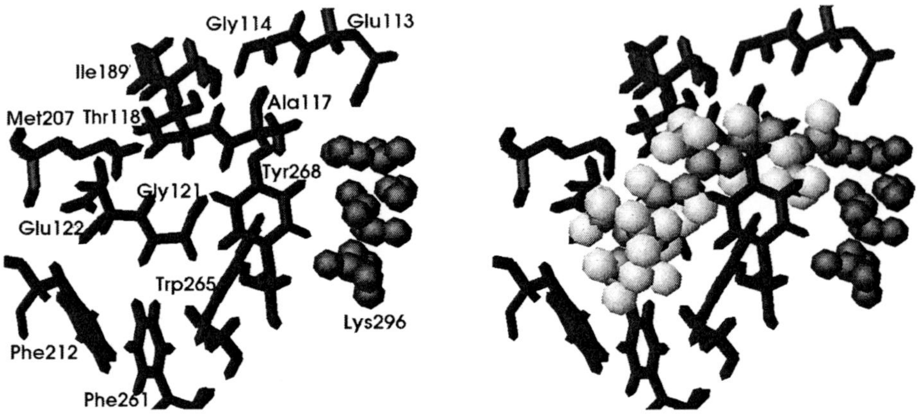

Figure 4. The schematic diagrams of a retinal binding pocket in the rhodopsin structure are shown. The chromophore retinal chain and Schiff base linkage are drawn by balls. The position of beta-ionone ring of the retinal polyene chain is left-hand side.

disruption of this binding site activates of rhodopsin at the absence of a retinal chromophore. In later stages of rhodopsin photolysis (META-I and META-II intermediates) the major conformational changes of rhodopsin are occured.

However, many uncertainties regarding the pathways of the rhodopsin conformation after photoexcitation still remain. Actually the current scheme of rhodopsin photolysis is that as it is represented above. Nevertheless, some papers discuss the existence of two or even three forms of rhodopsin intermediates on each stage of photolysis [8, 9, 10, 11, 12]. By other words, a multiple conformation state of visual pigment rhodopsin on different stages of its photolysis is quite possible to exist. The recent NMR spectroscopic data and Density functional theory calculations had shown the existence of two conformational forms of beta-ionone ring relative to the chromophore (6-s-*cis*-11-*cis*-retinal and 6-s-*trans*-11-*cis*-retinal), which populate protein pocket in the native dark state of rhodopsin [13, 14]. Perhaps, the above conformational states may generate different interactions with rhodopsin helices, leading to multiple intermediate states on each stage of rhodopsin photolysis. Thus, we had applied a complex approach to investigate the primary stages of rhodopsin photolysis, using the low-temperature technique and the rhodopsin dark-adapted state through the Molecular Dynamics Simulations.

Part1: MD Simulations on Rhodopsin Structural Conformation

Simulation Details

We simulated the bovine rhodopsin in the dark-adapted state. The dynamics of rhodopsin conformation for dark-adapted state we examined through the considerations: (1) rhodopsin without retinal; (2) rhodopsin with 11-*cis* retinal. A molecular model was constructed from the refined rhodopsin crystal structure as it shown in Fig. 1. In the MD simulations we use only one chain (molecule A) from the possible twin structure [36, 37]. The missing amino acids 236-240 and 331-333 were inserted using the MOE software package for biomolecular simulations [62]. The molecular mechanics (MM) and MD simulations on rhodopsin were performed with an explicit water solvent (see Fig. 5). Two set of MD simulations (viz., cutoff and no cutoff calculations) were performed with the program package AMBER [28, 63].

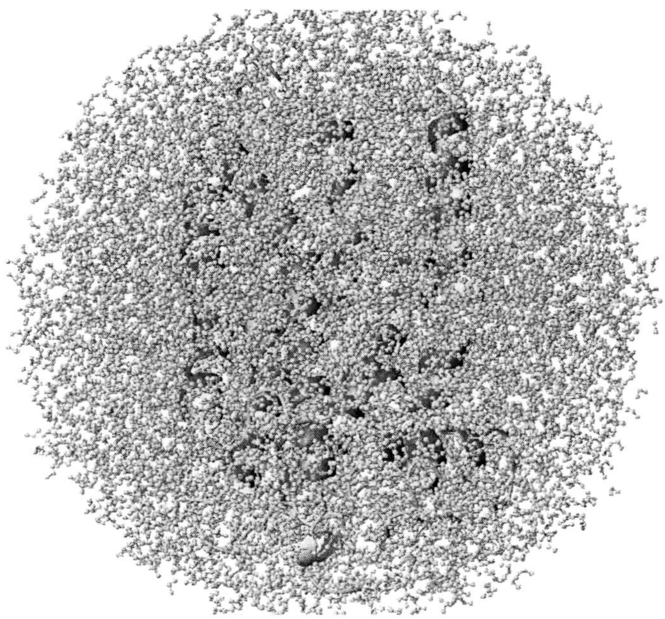

Figure 5. The equilibrium configuration of rhodopsin protein is shown. The molecular mechanics and molecular dynamics simulations of rhodopsin were performed with a water solvent (the solvent TIP3P water molecules are drawn by small balls).

We have used the program package AMBER 5.0 (Parm94, cutoff method) [63] and modified version of AMBER7.0 (Parm96, no cutoff method) for a special-purpose computer MDGRAPE-2 [64, 65, 66]. The all-atom force field of Cornell et al. [67] was used in the MD simulations. A system was solvated with TIP3P molecules [68] generated in a spherical (non-periodic) water bath. The system temperature was kept constant by using the Berendsen algorithm with coupling time of 0.2 ps [69]. Only bond lengths involving hydrogen atoms were constrained using the SHAKE method [70]. The cutoff distance of non-bonded interactions was 14 Å; for the MDGRAPE-2 all interactions were calculated. The integration time step in the MD simulations was 1 fs and the rhodopsin dynamics was studied for the $3 \cdot 10^6$ steps (3 ns state). The simulation procedures were the same in all calculations [51, 50]. Firstly, for each system a potential energy minimization was performed on an initial state. Next, the MD simulation was performed on the energy-minimized states. The temperatures of the considered systems were gradually heated to 300 K for and then kept at 300 K for the next 3 million time steps. The trajectories at 300 K for 3 ns were compared and studied in detail. The result of simulations and images of simulated proteins were analyzed by RasMol [71] and MOLMOL [72] packages.

Simulation Results

A. Rhodopsin without Retinal

In this consideration all kind interactions inside the rhodopsin pocket and the solvent environment for the chromophore retinal were neglected. Firstly, we present the results for the rhodopsin RMSD (root-mean-square displacement) changes. In Figs. 6 and 7 the RMSD for a total rhodopsin structure domain and for each of the helices I-VII are shown, respectively.

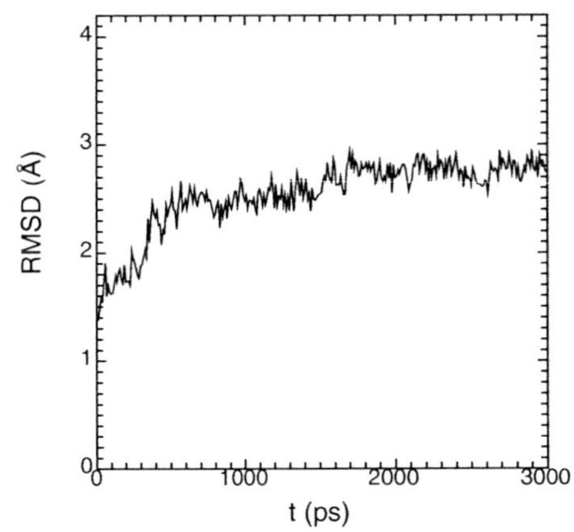

Figure 6. The RMSD for the rhodopsin total structure is shown.

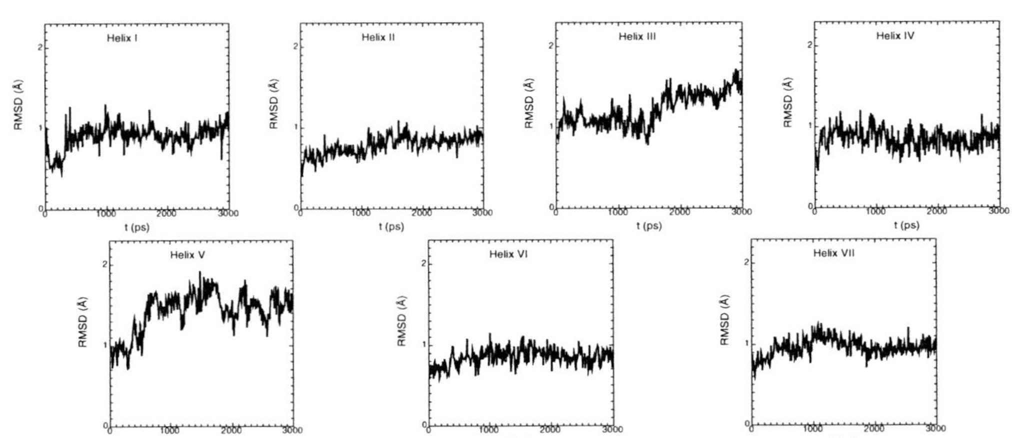

Figure 7. The RMSD values calculated for each of rhodopsin helices I-VII are presented. Helices I, III and V show highest deviations from the reference structure.

The RMSD values were evaluated for only the backbone atoms of the helices. The RMSD of the total rhodopsin structure, as it is seen from Fig. 6, has kept around 4.5 Å, indicating on a structural stability during 3.0 ns simulations. With regard to a selected helix (Fig. 7) we observe a relative movement of helices; displacement of the rhodopsin helices have different behavior. Helices I, III and V exhibit the highest deviations from the reference structure.

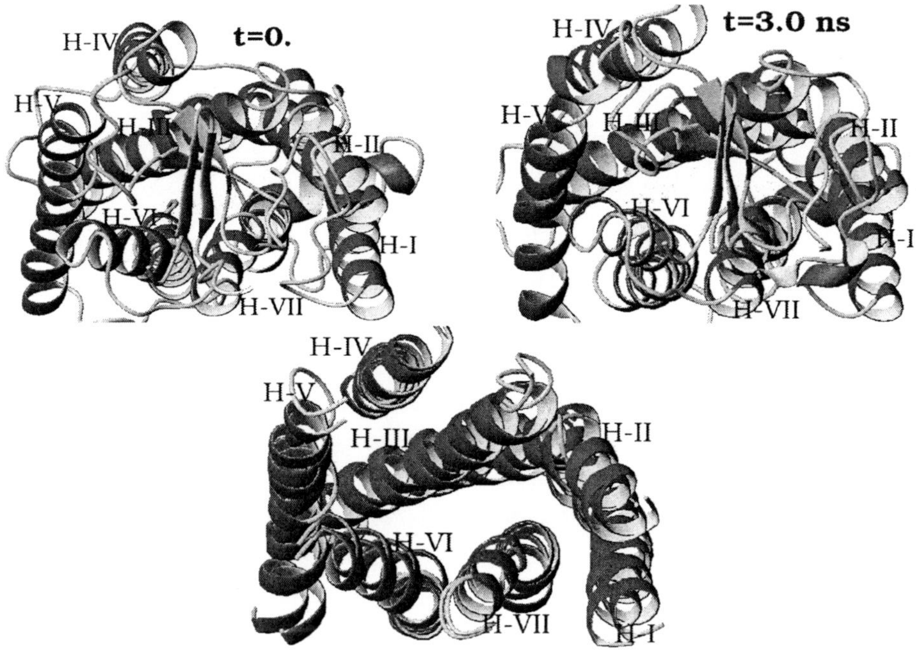

Figure 8. A. (top) The initial and final (3 ns) rhodopsin structures are shown. B. (bottom) The superposition of the representative structures are displayed. A largest displacement from the reference structure are seen for the helices III and V, respectively.

In Fig. 8 the superposition of the representative structures is presented at 3 ns state. Fig. 8 shows the largest displacements from the reference structure for the helices III and V. The largest differences are seen in the cytoplasmic region of the helices; the extracellular ends show slight movement. Some reported MD results indicate, for example, the largest RMSDs for helices I, V and VI or helices IV, V and VI [34, 59, 60]. It is worth noting that the spin-label experiments show the movement of the cytoplasmic part for the helices II, VI and VII [26, 43, 61].

B. Rhodopsin with 11-*cis* Retinal

Next, we simulated the rhodopsin conformation dynamics with the inclusion of the 11-*cis* retinal at the start of runs. In Figs. 9 and 10 the RMSD results for the rhodopsin total structure domain and for each rhodopsin helices are presented.

As it is seen from Fig. 9, the inclusion of a small retinal chain into the whole dynamics does not influence the RMSD for the total rhodopsin structure. The RMSD during the 3 ns dynamics possesses the same average value 4.5 Å, comparing Figs. 9 and 6, and the RMSD has the same behavior as in the absence of the retinal chromophore. Nevertheless, with regard to a selected helix, the inclusion of the retinal chromophore results for all the RMSD values to change. From Figs. 10 the RMSD, separately estimated for each of helices, has clearly different behavior between the rhodopsin with 11-*cis* retinal and rhodopsin without retinal configurations. The comparison of the RMSD values are straightforward (see Figs. 10 and 7).

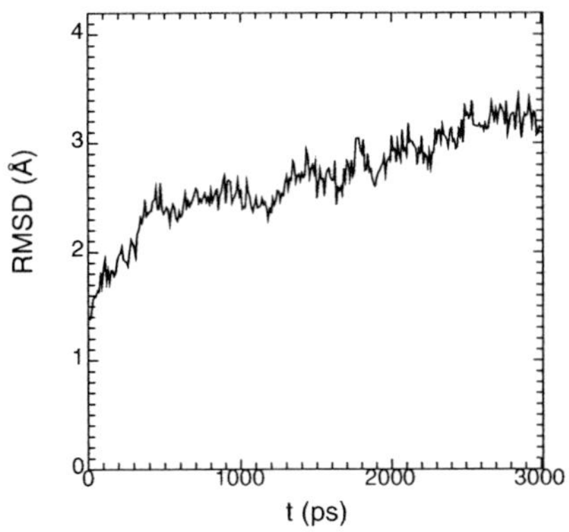

Figure 9. The RMSD (root-mean-square displacement) of the total rhodopsin structure is shown. (The retinal in 11-*cis* conformation at the start of runs, all atomic interactions with rhodopsin and solvent.)

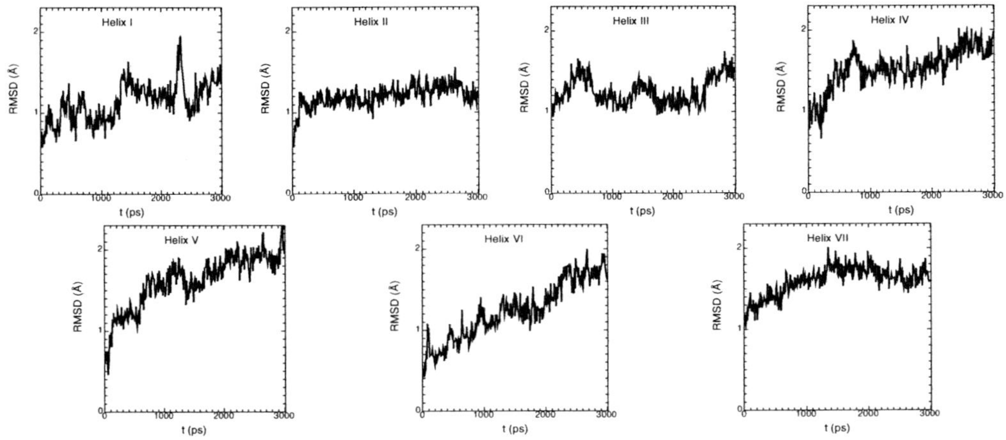

Figure 10. The RMSD values calculated for each of helices I-VII are presented. (The retinal in 11-*cis* conformation at the start of runs, all atomic interactions in rhodopsin and solvent.)

C. Rhodopsin Conformation in the Dark-adapted State

In Figs. 11 four sequential snapshots of the MD simulation are presented for rhodopsin ribbon structure with a retinal chromophore. The dynamical changes of the retinal chain inside the rhodopsin pocket are drawn separately with the balls.

From Figs .11 we observe the rhodopsin helical deviations for both structural regions, viz., beta-ionone ring (left-hand side) and the Schiff base linkage (right-hand side). In comparison to the initial state (right pictures) the visibly high helical distortions have seen at later stages (left pictures) within the frame of the 3 ns period. It is remarkable to note that rhodopsin helical distortions, as it is seen from the Figs .11, have to be accompanied with a rotation of the beta-ionone ring.

In Fig. 12 the positions of the key amino acids within the binding pocket surrounding the retinal chain are shown. As it seen from Fig. 12 the retinal polyene chain during the dynamical changes inside the rhodopsin pocket has to be transformed in the beta-ionone

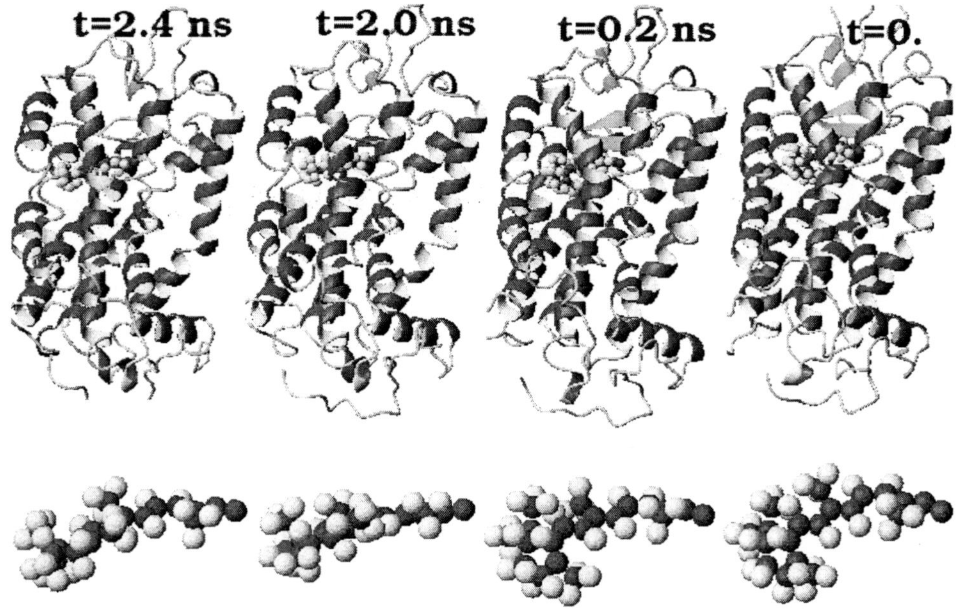

Figure 11. Four sequential snapshots of the MD simulation. The rhodopsin ribbon structure and with retinal chromophore are shown (The dynamical change of the retinal chain are separately presented by the balls). The retinal polyene chain slightly rotates inside the rhodopsin pocket and it takes at final state (left picture, t=3 ns) a position perpendicular with regard to the initial state (right pictures, t=0).

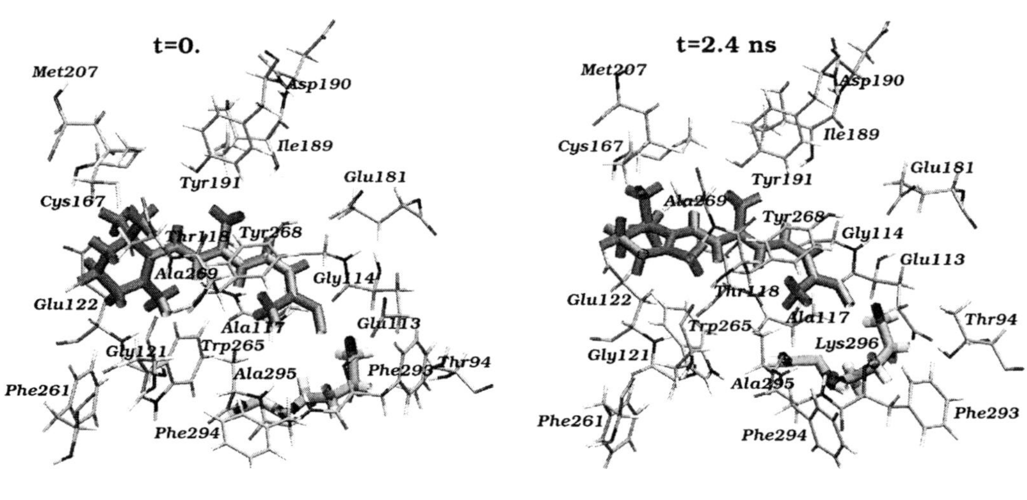

Figure 12. A schematic representation of the retinal binding pocket region is shown at the initial (t=0) and later (t=2.4 ns) states. The positions of the key amino acids surrounding the retinal chain are drawn within the binding pocket. The retinal chain slightly transforms relatively to the initial conformation.

region. The beta-ionone ring of the retinal chromophore at later stages (that is in frame of 3 ns period) takes a position perpendicular with regard to that at the initial state.

D. Rotational Changes of the Beta-ionone Ring

Let us look on the details of the atomistic pictures generated for rhodopsin and retinal conformation inside its binding pocket. In Figs. 13 we present the calculated rhodopsin structures at the initial and final (3 ns) states (shown from the cytoplasmic part). In Figs. 13 we observe that the structural changes occur for all helices and in both parts of the rhodopsin protein (viz., the Schiff base linkage: right-hand side, and beta-ionone ring: left-hand side). The visibly significant conformational changes have been observed in the beta-ionone region. It is easy to see that the beta-ionone ring in course of its dynamics takes a more planar form relatively to the helix III. Also the helix VI in the final state reaches a more close position to the retinal chromophore. To enlighten the mechanisms underlying these conformational changes, a number of dynamical movies for the rhodopsin + retinal chromophore + water solvent were generated; the different structural positions and views were compared. The structural observations are summarized in the following three sets of the pictures (Figs. 14-16).

In Figs. 14 we see that the rotational changes of the beta-ionone region originates at the time [0.3-0.4] ns from the start of the simulations (side view: the same as in Figs. 11) Within this time interval we observe the orientational changes of the amino acid residue Phe-261 (H-VI): this residue changes its position to an opposite site of the retinal chain. From our simulation we have found that the orientation of the residue Phe-261 occurs within a short interval [0.3-0.4] ns and that this is clearly accompanied by the rotation of the beta-ionone ring.

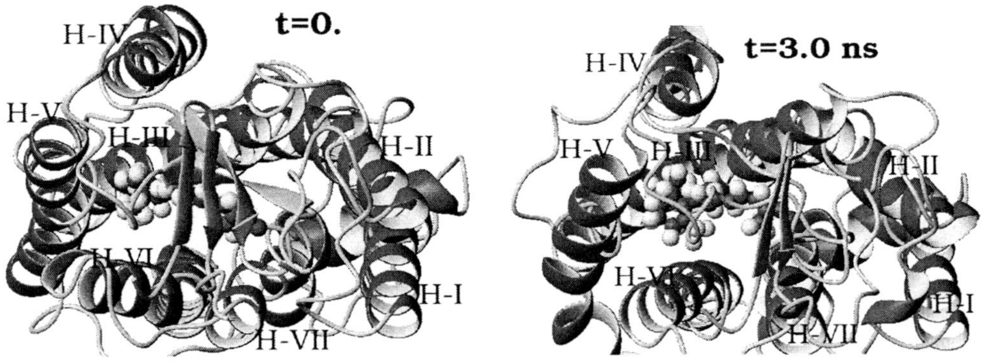

Figure 13. The initial and final (t=3 ns) rhodopsin ribbon structures are presented (view are shown from the cytoplasmic side). The retinal chain is drawn by balls. The structure of the beta-ionone ring in comparison with the initial state became a more planar with regard to the helix III of the protein. At the final state helix VI reaches a more close position to the retinal chromophore.

Starting from this point (at [0.3-0.4] ns) the beta-ionone ring kept its conformation almost planar with regards to the helix III (until the end of simulations, for about 2,6 million MD-time steps!). Thus, the Phe-261 orientational change may probably initiates the retinal

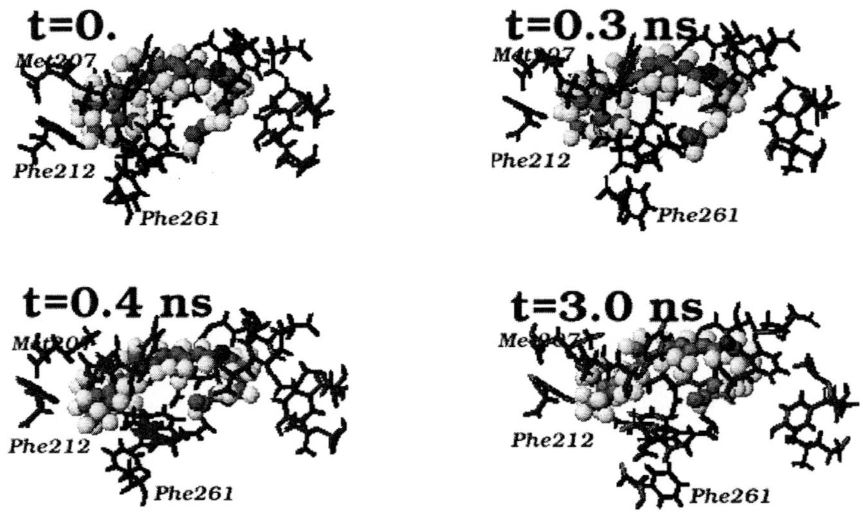

Figure 14. The conformation of the retinal chain and positions of the key amino acids inside the rhodopsin binding pocket are shown (side view, the same like in Figs. 11). The rotational changes of the beta-ionone region originates at the interval [0.3-0.4] ns. Simultaneously within this temporal interval residue Phe-261 changes its position in relation with the retinal chain to an opposite site. The orientational change of the residue Phe-261 seen to be accompanied with the rotation of the beta-ionone region.

conformation change inside its binding pocket. However, in formation of the retinal conformation may be involved other amino acid residues or atomic sites. It is worth to noting that retinal binding pocket is large and it is enough flexible which can accommodate various chromophores.

In Figs. 15 we present a competition of the two amino acid residues (viz., Phe-212 and Ala-269) to be accommodated around the beta-ionone region. (view from cytoplasmic part: the same view as in Figs. 13). Until the temporal point [0.3-0.4] ns (say, a key point for the rotation of the beta-ionone ring) residues Phe-212 and Ala-269 are equidistantly positioned with regard to the polyene part of the chromophore. As rotation of the beta-ionone ring happens residue Ala-269 reach a more close distant to the retinal (in beta-ionone region) than that Phe-212. This perhaps results in the total helix VI displacement (to which the residue Ala-269 belongs) to be positioned more close to the chromophore retinal (in beta-ionone part).

In Figs. 16 a summary of the structural changes of the only retinal chromophore at the different moments of time within 3-ns simulation are presented. The pictures in Figs. 16 display the retinal conformations for the same structural positions as described above (left: side view, the same as in Figs. 11 or 14; right: view from the cytoplasmic part, the same as in Figs. 13 or 15).

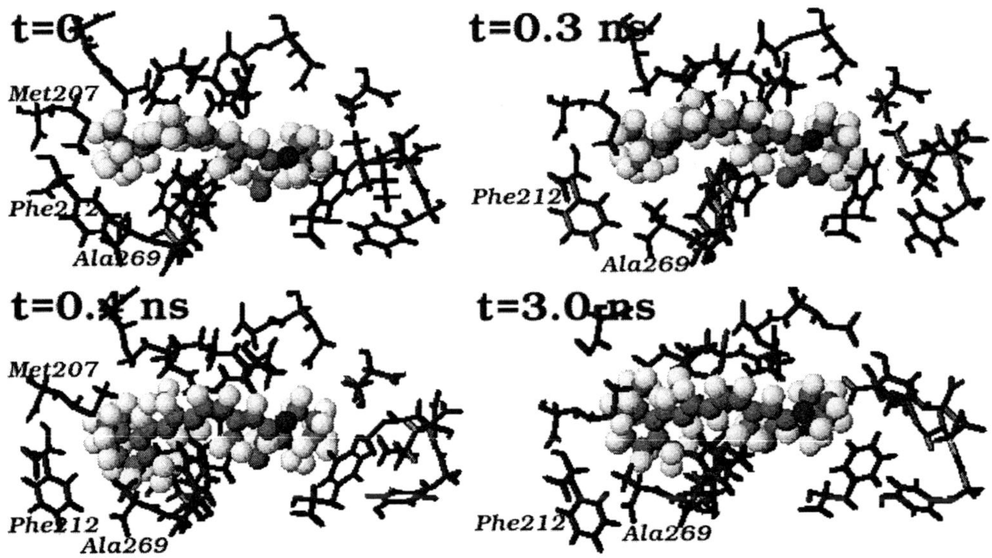

Figure 15. The conformation of the retinal chain and key amino acid positions inside the rhodopsin binding pocket are shown (view from the cytoplasmic part of rhodopsin protein, the same as in Figs. 13). The positions of two amino acid residues, viz. Phe-212 and Ala-269, are compared. With the rotational changes of the beta-ionone ring at the interval [0.3-0.4] ns the residue Ala-269 takes a more close position to the retinal site than that Phe-212.

E. The Details of the Rhodopsin Helical Movement

In this section we analyze rhodopsin conformational changes from the point of view of its interhelical movement. With regards to the structural conformation, the inclusion of a small retinal chain does not affect the RMSD average value. In other words, a small retinal chromophore, inside the rhodopsin and the embedded solvent environment, does not change the picture of atomic interactions. The RMSD, as it is seen from Fig. 17, for the total rhodopsin structure has almost the same average value for both the rhodopsin with 11-*cis* retinal (solid line) and rhodopsin without retinal (dotted line). From this comparison it is easy to observe that the RMSD in both consideration have the same dynamical behavior.

It turns out that the inclusion of a retinal chromophore into the rhodopsin dynamics results in essential interhelical movement and conformational changes. Based on the RMSD analysis, we have estimated the helical changes for the rhodopsin with 11-*cis* retinal and rhodopsin without retinal. We consider the rhodopsin interhelical conformation in the following two regions: (1) the Schiff base linkage and (2) the beta-ionone ring.

E.1. The Schiff Base Linkage Region

The Schiff base linkage region for the retinal chromophore includes the helices I, II and VII. In Figs. 18-20 the structural conformations of these helices along with the retinal positions are displayed. From Figs. 18 we observe that the RMSD for the helix I (solid line for rhodopsin with 11-*cis* retinal) follows a nonlinear dynamics. Starting with a monotonic

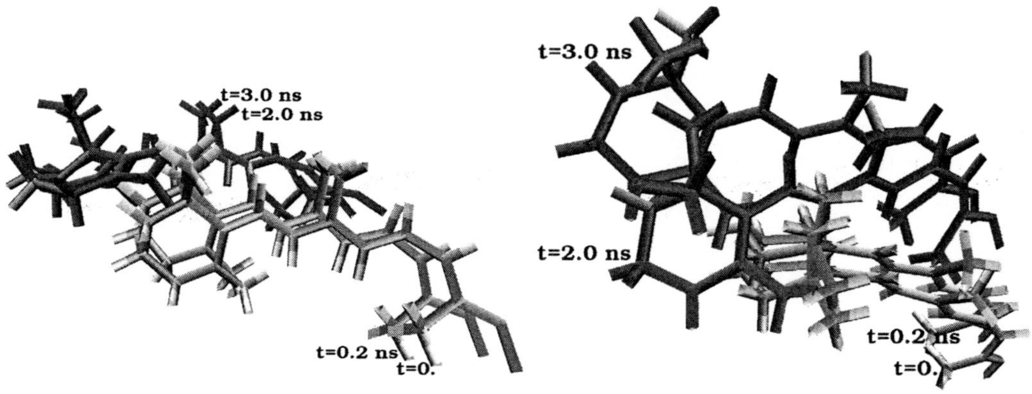

Figure 16. The structural changes for the only retinal chromophore at the different moments of time are presented. The pictures display the retinal conformations on the same structural positions above (left: side view, the same as in Figs. 11 or 14; right: view from the cytoplasmic part, the same as in Figs. 13 or 15).

change at t=0, the behavior of the RMSD becomes non-monotonic with time. Its value jumps around the points t=1.5 ns and 2.5 ns, indicating the temporal distortion of this helix. As it is seen in Figs. 18, the distance between helix I and retinal chain (in polyene part) gradually increases. The rotational change of the beta-ionone ring is directed normally to the axes of the helix I (bottom snapshot). In Figs. 19 the comparison of the RMSD, the distance and the structures for the helix II, are displayed. Helix II for the rhodopsin with 11-*cis* retinal have a higher deviation in comparison to the rhodopsin without retinal configuration. The positions of the helix II and the polyene retinal chain changes monotonically and that the helical distortions are not so large. In Figs. 20 the comparison of the RMSD, the distance and the structural conformations of the helix VII display visibly high deviations. To summarize the rhodopsin structural behavior in the Schiff base linkage region, we conclude that the retinal inclusion causes in this part of the protein the largest helical distortions. It is worth noting that all the rhodopsin structural changes described above occur within the retinal 11-*cis* configuration.

E.2. The Beta-ionone Ring Region

The beta-ionone region of the retinal chromophore includes the rhodopsin helices III-VI. The conformations of these helices along with the position of the retinal chain (in beta-ionone region) are presented in Figs. 21-24. For the helix III (Figs. 21) the inclusion of the retinal chromophore seems helpful to maintain its original shape. Comparison of the RMSD (solid and dotted lines) shows that the helix III does not deviate from its free form (at the absence of the retinal). The dynamical orientation of the retinal chain with time relatively to the helix III becomes planar. Such rotational change of the retinal chain seems to play a role in the structural stabilization of the helix. For the helices IV and V (Figs. 22 and 23) the RMSD have greater values, if compared to the free form (solid and dotted lines). The distances from these helices to the retinal site (beta-ionone part) become shorter. In comparison to the initial state the helices IV and V accommodate in the beta-ionone region more close positions to the retinal chromophore. In Figs. 24 the conformation behavior of the helix VI is shown. The RMSD of the helix VI increases within a simulation time. The

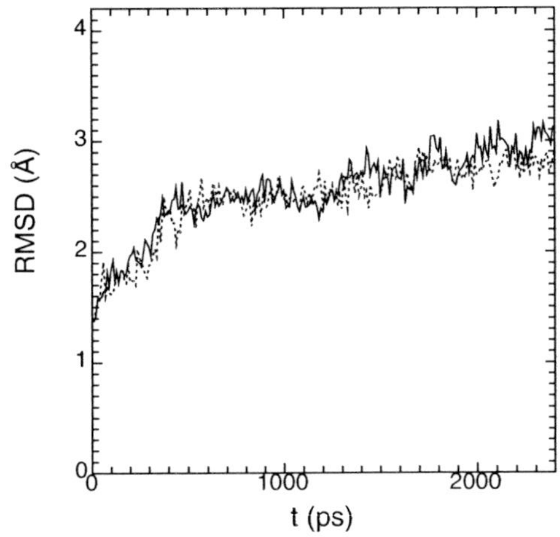

Figure 17. The RMSD evolutions of the total structure for the rhodopsin with 11- *cis* retinal (solid curve) and rhodopsin without retinal (dotted curve).

increasing of the RMSD, however, does not accompanied with the distortion of the helix VI outside of the rhodopsin binding pocket. On contrary, helix VI takes a more close position to the beta-ionone ring of the retinal chain. To summarize, the rhodopsin conformation in the beta-ionone region has to be characterized by the less helical distortions than that in the Schiff base linkage region.

Part 2: Low-temperature Spectroscopy Studies of Rhodopsin

Spectral Measurements Procedures

Rhodopsin Preparation

All the procedures were performed under dim red light. Digitonin extracts of rhodopsin obtained from rod outer segments (ROS) suspension of bovine retina has been used [15]. The purity of obtained suspension was equal $A_{500\ nm/280\ nm} = 2{,}0\text{-}2{,}1$. The ROS suspension was solubilized by 2% digitonin in Na^+- phosphate buffer (0.1 , 7.4) contained 1 mM gCl_2, 0,1 m DTA and 0.1 hydroxylamine. All samples contained 65% glycerol.

Spectral Measurements of Rhodopsin Transition at -155 ° and Above

The rhodopsin samples were irradiated at -155 ° by blue light at a wavelength of 436 nm as described in [12]. The sample was placed in a specially designed quartz Dewar flask [16]. Irradiation was performed with a DRSh-250 mercury lamp, using Hg 436 and 10 cm water filters. Steady-state absorption spectra were measured using a Shimadzu UV-1601PC spectrophotometer (Japan).

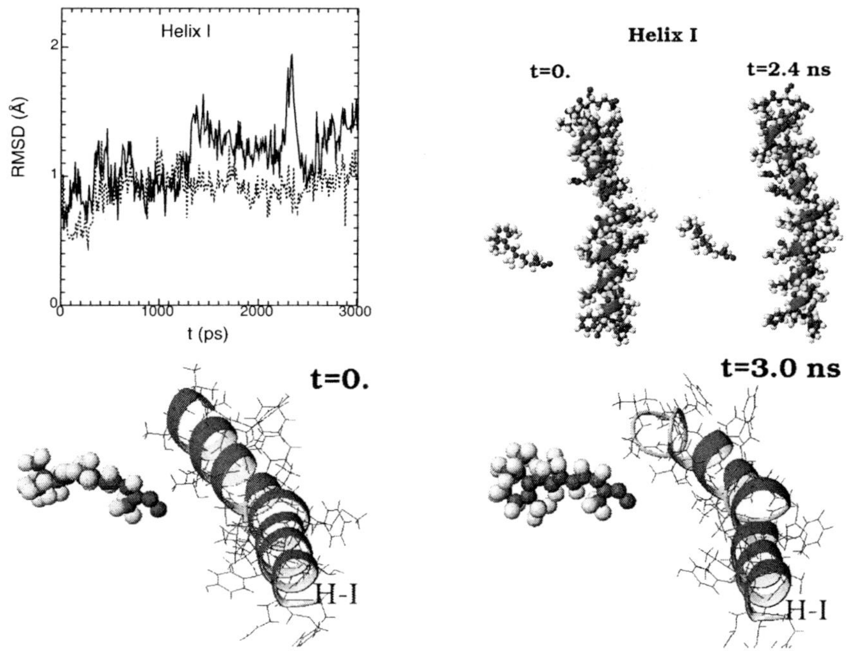

Figure 18. The RMSD values of the helix I (solid line: rhodopsin with 11-*cis* retinal; dotted line: rhodopsin without retinal) are displayed along with the positions of the helix in relation with the retinal chromophore (top snapshot: the distance view; bottom: view from the cytoplasmic side).

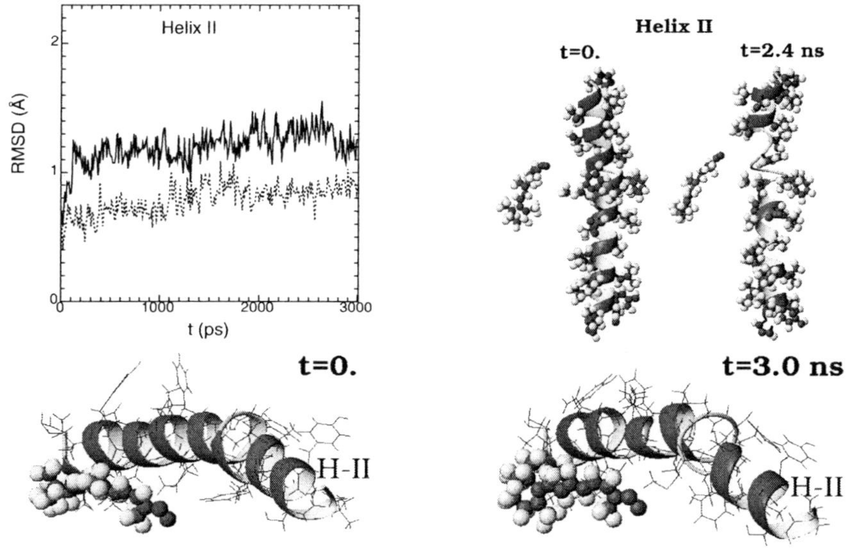

Figure 19. The RMSD values of the helix II (solid line: rhodopsin with 11-*cis* retinal; dotted line: rhodopsin without retinal) are displayed along with the positions of the helix in relation with the retinal chromophore (top snapshot: the distance view; bottom: view from the cytoplasmic side).

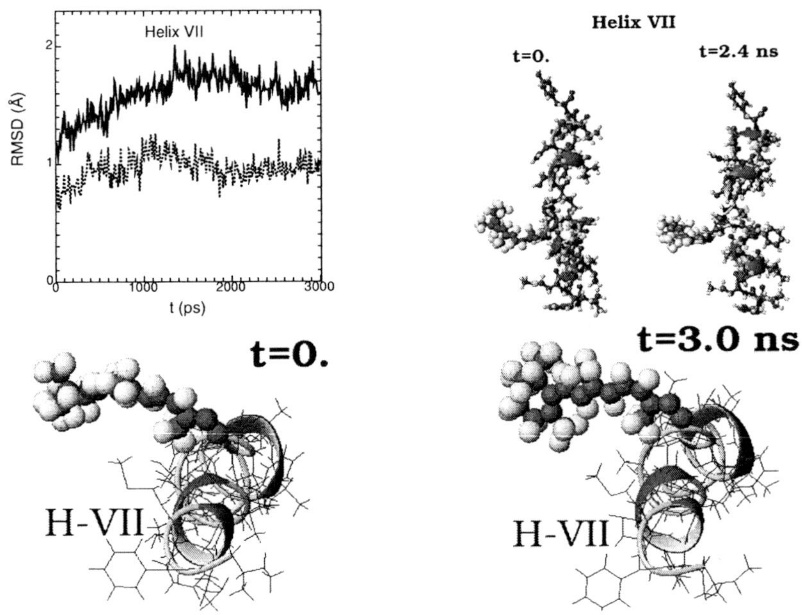

Figure 20. The RMSD values of the helix VII (solid line: rhodopsin with 11-*cis* retinal; dotted line: rhodopsin without retinal) are displayed along with the positions of the helix in relation with the retinal chromophore (top snapshot: the distance view; bottom: view from the cytoplasmic side).

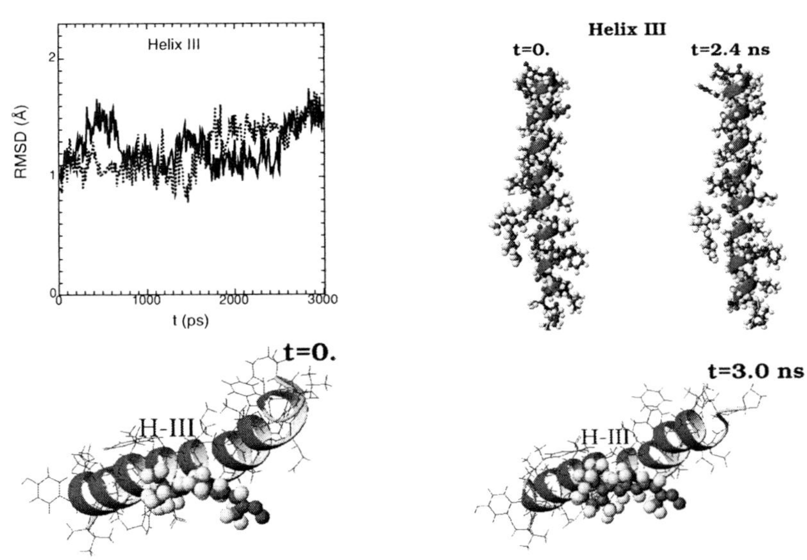

Figure 21. The RMSD values of the helix III (solid line: rhodopsin with 11-*cis* retinal; dotted line: rhodopsin without retinal) are displayed along with the positions of the helix in relation with the retinal chromophore (top snapshot: the distance view; bottom: view from the cytoplasmic side).

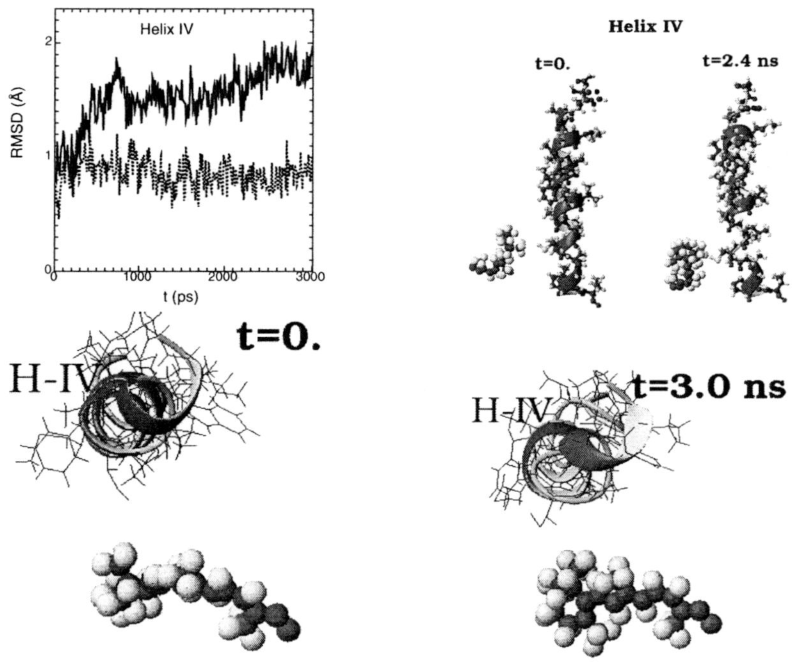

Figure 22. The RMSD values of the helix IV (solid line: rhodopsin with 11-*cis* retinal; dotted line: rhodopsin without retinal) are displayed along with the positions of the helix in relation with the retinal chromophore (top snapshot: the distance view; bottom: view from the cytoplasmic side).

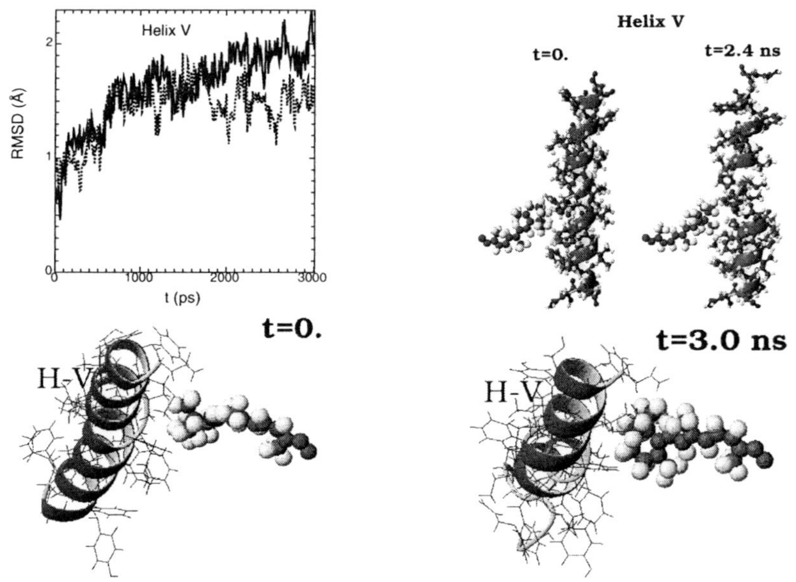

Figure 23. The RMSD values of the helix V (solid line: rhodopsin with 11-*cis* retinal; dotted line: rhodopsin without retinal) are displayed along with the positions of the helix in relation with the retinal chromophore (top snapshot: the distance view; bottom: view from the cytoplasmic side).

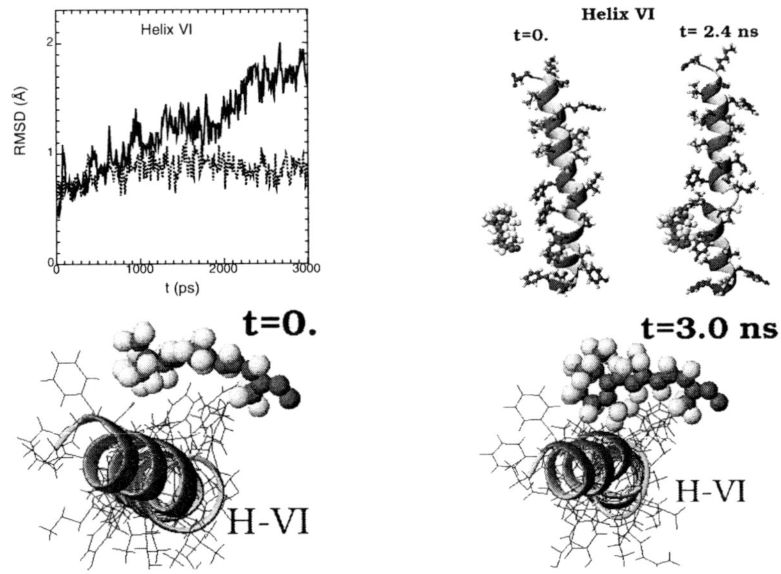

Figure 24. The RMSD values of the helix VI (solid line: rhodopsin with 11-*cis* retinal; dotted line: rhodopsin without retinal) are displayed along with the positions of the helix in relation with the retinal chromophore (top snapshot: the distance view; bottom: view from the cytoplasmic side).

Low-Temperature Spectroscopy Results

The rhodopsin was irradiated at -155 °C by blue light of the wavelength 436 nm. As it is known the bathorhodopsin is stable at this temperature [3]. As a result of irradiation the steady state mixture of rhodopsin and bathorhodopsin is obtained. This mixture is contained near 70% of product with all-*trans* retinal [11]. The next, classical product of rhodopsin photolysis lumirhodopsin is appeared after raising of temperature to near -90 ° [17]. Both absolute and differential spectra that show this photoconversion are presented on the Fig. 25.

However, if we have increased the temperature very slow and by step 1-3 °C and recorded the spectra in the dark, we were able to observe during batho- to lumirhodopsin transition a simultaneous appearance more than one spectral products (Fig. 26). One can see a rise of absorption around 440 nm at temperature between -155 and -145 ° (Fig. 26, top: spectrum 1-3; bottom: spectrum 1-4). It is can not be lumirhodopsin, because absorption maximum of differential spectrum of lumirhodopsin situated around 475 nm (Fig. 25, bottom: spectrum 2). We will call this product as D_{440} (D means absorption maximum or minimum on differential spectrum).

Also, one can see a decrease of absorption minimum around 520 nm in the course of temperature rise. It means that an intermediate product with minimum of absorption at 520 nm on the differential spectrum disappears. This D_{520} nm product is not a bathorhodopsin, since its minimum of absorption on the differential spectrum situated at 555 nm (Fig 25, bottom: spectrum 2). In fact, this D_{520} nm product is a precursor of the other product with maximum of absorption at 440 nm. Indeed, one can see the isobestique point at 460 nm in the course of the D_{520} nm product transition to the D_{440} nm product.

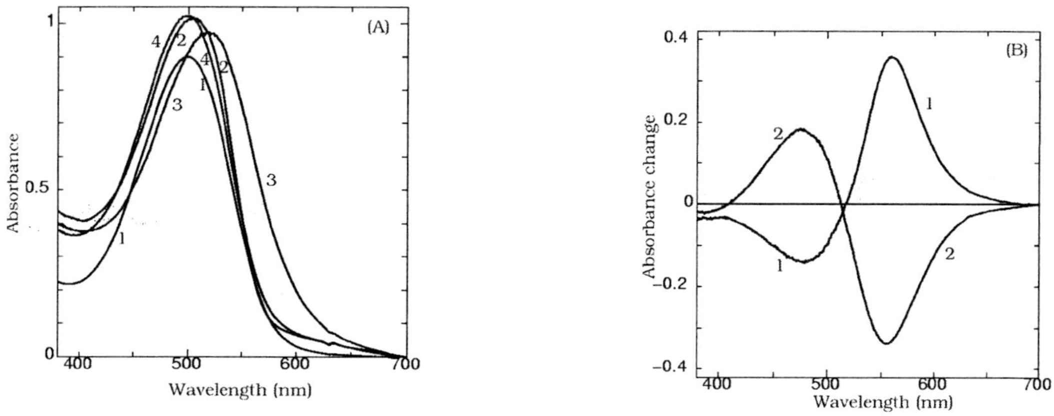

Figure 25. Absorption spectra of rhodopsin and its photointermediates. A. Absolute absorption spectra. 1 dark rhodopsin at 22 °C; 2 dark rhodopsin at -155 °C; 3 irradiated rhodopsin at -155 °C, 436 nm, 5 min (photostationary mixture of rhodopsin and bathorhodopsin); 4 warmed photostationary mixture from -155 °C to -115 °C (stationary mixture of rhodopsin and lumirhodopsin); B. Difference spectra. 1 rhodopsin to bathorhodopsin transition (A: curve 3 minus curve 2); 2 bathorhodopsin to lumirhodopsin transition (A: curve 4 minus curve 3).

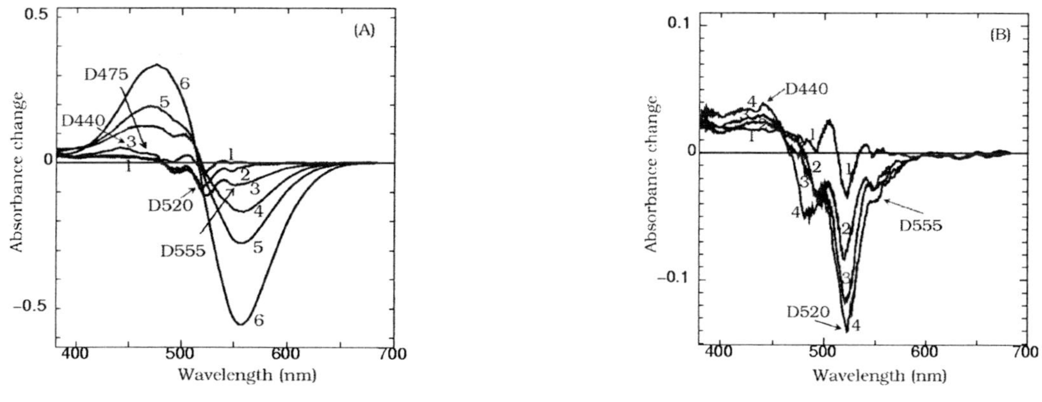

Figure 26. Difference absorption spectra due to the thermal reaction of rhodopsin intermediates (after irradiation of rhodopsin at -155 °C, 436 nm, 5 min) are calculated by subtracting the absolute absorption spectrum (Fig.1, curve 3) recorded immediately after irradiation from the spectra recorded at later times after the irradiation in the course of a slow rise of temperature. A. Difference absorption spectra recorded in the temperature interval from -155 °C to -115 °C (spectra 1-6). 1 at -155 °C in 5 min after irradiation; 2 at -145 °C in 15 min after irradiation; 3 at -130 °C in 1h after irradiation; 4 at -130 °C in 2 h after irradiation; 5 at -120 °C in 3 h after irradiation; 6 at -115 °C in 3,5 h after irradiation. B. Difference absorption spectra recorded in the temperature interval from -155 to -145 °C. 1 at -155 °C in 5 min after irradiation; 2 at -145 °C in 15 min after irradiation; 3 at -145 °C in 20 min after irradiation; 4 at -145 °C in 25 min after irradiation.

So, the existing of isobestique point supports our point of view that the D $_{440}$ nm product is formed from the D_{520} nm product. Further temperature rise leads to formation of new products as it can see from the appearance of absorption bands on the differential spectra at 475 555 nm. We suppose that these changes reflect the classical transition from batho- to lumirhodopsin. One should stress that in this case (at this higher temperature) the isobestique point is absent. It indicates again that the batho- to lumirhodopsin transition is not direct but more complicate process.

We can suppose that $D_{520} \longrightarrow D_{440}$ and $D_{555} \longrightarrow D_{475}$ transitions are occurred actually simultaneously (in parallel), and the first one - $D_{520} \longrightarrow D_{440}$ is ahead of the second one - $D_{555} \longrightarrow D_{475}$.

The differential spectra contain at temperatures from -110 ° and higher only one maximum at 475 nm and one minimum at 555 nm. The shape of these spectra becomes classical under such conditions, and it described in literature as a batho- to lumirhodopsin transition [3, 17].

Based on the differential spectra analysis one can conclude that disappearance of D_{520} and appearance of D_{440} is not actually bathorhodopsin $\longrightarrow$ lumirhodopsin transition - ($D_{555} \longrightarrow D_{475}$) or bathorhodopsin $\longrightarrow$ BSI ($D_{555} \longrightarrow D_{440}$). It is observed the appearance more than one, at least two, photointermediates on these early stages of rhodopsin photolysis. One can suggest the existence of parallel ways of rhodopsin photolysis.

Discussions

In present work based on the MD simulation and spectral studies we have investigated the rhodopsin molecule dynamical behavior and conformation properties. We have performed 3 ns MD-simulations on the rhodopsin and presented the structure analysis for the dark-adapted state. We have compared the dynamical properties of the rhodopsin protein in the following considerations: (1) Rhodopsin without retinal; (2) Rhodopsin with 11- *cis* retinal. In the former approach for the retinal chromophore all kind interactions inside the rhodopsin pocket and with water solvent were neglected. The analysis of the RMSD (root-mean-square deviation) and generated structure images for the rhodopsin (without retinal) show a relative movement of rhodopsin helices, that possesses different behavior between the helices. Helices III and V of the rhodopsin exhibit the highest deviations from the reference structure. In the latter consideration for the rhodopsin (with 11-*cis* retinal) we generated the pictures of the atomic-scale processes for the binding pocket, surrounding the chromophore retinal, compared the helical deviations in the beta-ionone ring and Schiff base linkage regions of the protein. We have observed that the inclusion of small retinal chain into the whole rhodopsin dynamics (+ solvent) does not affect the RMSD behavior for the total rhodopsin structure. Within the 3-ns time step calculations the RMSD of the whole rhodopsin molecule (11-*cis* retinal) possesses the same average value and behavior as in the absence of the retinal chromophore. However, one should stress out that concerning a selected rhodopsin helix, the inclusion of the retinal changes all RMSD values and causes essential helical distortions.

The structural images were generated to display the conformation of each of the rhodopsin helices. The detailed structural configurations were separately estimated for the beta-ionone ring and Schiff base linkage regions of rhodopsin. The Schiff base linkage region for the retinal chromophore includes the helices I, II and VII. The RMSD of helix I shows a nonlinear dynamics, indicating the temporal distortion of the helix: the distance between the helix I and the retinal chain gradually increases. The temporal positions of the helix II and the retinal chain show that the helical distortion to be not so large. The RMSD, distance and structural data of the helix VII indicate a visibly high deviation. The beta-ionone region of the retinal chromophore involves the helices III-VI of the rhodopsin. The inclusion of the retinal chromophore for helix III seems helpful to maintain its original shape. The comparison of the RMSD for the helix III shows that the helix does not deviate as much as similar in the absence of the retinal. The orientation of the retinal chain with regards to the helix III became a clearly planar, and that this rotational dynamics of the retinal chain probably plays an important role in the structural stabilization of the helix. For the helices IV and V the RMSD show significant higher deviations, however the distances from these helices to the retinal site (in beta-ionone region) seems to become shorter. In comparison to the initial state, helices IV and V accommodate a more close positions relatively to the retinal chromophore. RMSD calculated for the helix VI increases within a long period of simulation time. The increasing of the RMSD, however is not accompanied by the helical distortion; on the contrary, the helix VI accommodates a close position to the retinal chain. It is worth noting that some reported MD results indicate the largest RMSDs for helices I, V and VI or helices IV, V and VI [34, 59, 60]. The spin-label experiments, for example, show the movement of the cytoplasmic part for the helices II, VI and VII [26, 27, 43, 61].

It is remarkable that the rhodopsin helical distortions are accompanied by the rearrangement of the retinal chromophore, viz. with a tuning of the beta-ionone ring inside the protein binding pocket. From our simulation results we observed that within a short interval [0.3-0.4] ns the beta-ionone ring changes its orientation along the retinal chain. It is worth noting that such reorientation, however, does not occur for different parts of the retinal chromophore simultaneously. In accordance to our RMSD data and generated animations (structure movies), from the start of the above time interval the oscillation of a retinal methyl group C20 could easily be distinguished as oscillation of a pendulum perpendicularly to the polyene chain. That is the C20 methyl possesses a higher amplitude if to compare the latter to other retinal groups. It seems that C20 methyl group swings up the whole chromophore molecule, helping the chromophore to change an arrangement in its local site and to reach a more favorable state. As the retinal reorientation occurs the different groups of the retinal chain preserve their oscillation peculiarities. One can easily to observe the rotational type oscillations of the methyl groups C19 and C20 around the polyene chain (as oscillation of a pendulum), where C20 oscillates with a greater amplitude. At the same time the oscillation of these methyl groups are not synchronized. One can propose an assumption that these ocsillations may serve as a premise to or bringing to the strongly twisted retinal chain in the central region, from C10 to C13 [19, 20, 21, 22, 23], where this twist may be responsible for the initial ultrafast isomerization reaction. Moreover, as there have been reported in [24], after the retinal photoisomerization large 20 rotation and translation toward H-V disrupt the interactions of H-V, H-VI, and H-VII with the core of rhodopsin and lead to receptor activation [24]. It is very likely that for the rhodopsin dark-adapted state the large mobility of C20 group, in comparison to the other chromophore fragments, may serve as prerequisitive for a large C20 rotation in the excited-state rhodopsin. It is obvious that such dynamical behavior of the retinal chromophore inside the chromophore pocket has governed by the only protein surrounding.

It is worth to note that the experiments provide the data, for example, for the position of the beta-ionone ring, and that it is mostly covered from the cytoplasmic side by the residues of helices III and VI (the residues Glu-122, Phe-261, Trp-265) [30, 36]. The helix III provides many of the side chains for the binding pocket (say, Glu-113, Gly-114, Ala-117, Thr-118, Gly-120, Gly-121) surrounding the retinal polyene chain. From the extracellular side the side chains of Tyr-268 and Ile-189 determine the position of the C9-methyl of the retinylidene group. Side chains from H-V and H-VI (residues Met-207, His-211, Phe-212 Tyr-268 and Ala-269) also surround the beta-ionone ring [36, 52, 53, 54]. Moreover, the retinal reorientation may initiate, on the other hand, the conformational rearrangement of the amino acid residues in the chromphore location site. From our simulation results we observed that within a short interval [0.3-0.4] ns the beta-ionone ring changes its orientation in relation to the retinal chain. The beta-ionone ring at later stages takes more planar position with respect to the helix III of the rhodopsin. Within this interval the amino acid residue Phe-261 simultaneously changes its orientation relatively to the retinal chromophore. The beta-ionone ring rotation changes the positions of two amino acid residues, Phe-212 and Ala-269. Starting from [0.3-0.4] ns the residue Ala-269 accommodates close position to the chromophore retinal (in the beta-ionone site) than that a key residue Phe-212 does. Thus, our RMSD and other structural data confirm once more the importance of the chromophore-protein interactions. At the same time, our results are indicating on some molecular mechanisms of the protein and chromophore arrangement, under which the rhodopsin molecule may have its function or will behave as a highly sensitive sensor with the high quantum yeild (about 0.65) and isomerization speed (less 250 fs) [30, 36, 52, 53, 54].

Today the question about the truth retinal configuration in the chromophore site is one of the most arguing and discussed one. Unfortunately, the experimental structural (X-ray or NMR) data can not provide an unique answer here. Thus, at the present several theoretical and experimental work with contradictory conclusions have to exist. For example, there have been several reported results regarding on the beta-ionone ring conformation relatively to the retinal chain. Though it has been long assumed that the geometry is twisted 6-s-cis [21], a recent NMR [18] spectroscopic study has cast doubts on this and proposed that the conformation is derived from 6-s-trans instead. Moreover, it was discussed the existence of two conformations of the beta-ionone ring simultaneously. Density functional theory calculations in [14] assumed that protein pocket tolerates both conformations, 6-s-*cis* and 6-s-*trans*, where 6-s-*cis* is a more stable conformer than 6-s-*trans* and compound 3:1 equilibrium population at room temperature. The experimental study confirmed this prediction [13]; the authors based of NMR spectroscopic data concluded that the binding pocket is populated by two conformational states. A minor component comprising about 26 % and corresponding possibly to a twisted 6-s- *trans* form and a major component which the authors identify as the 6-s-*cis* conformer. Moreover, in [22] on the basis of the PM3-SDCI calculations the authors suggest that 6-s-trans geometry is preferred in the binding pocket under conditions when interactions with the protein environment are included. In recent studies [23], starting with models built from the X-ray data of the 2.6 Å structure [37] and their own of the 2.2 Å structure, QM/MM MD simulations have been performed. The obtained result supports the 6-s-cis form with substantial negative twist.

From our data presented above we also conclude that at the 3-ns MD calculated state 6-s-*cis* seems to be the most preferable conformer. At the same time this conformational state of rhodopsin (with a nonplanar distorted and twisted retinal) differs from that we have used [36] as initial state at the start of the MD simulations. In this rhodopsin state we have a retinal with a planar structure, like free 6-s-cis-11-cis-retinal conformation. In spite of the fact that our theoretical calculations are only model, we can suppose that the initial conformational state of the retinal may make a physiological sense. It is well known that the free retinal in solution usually takes a similar to planar structure. Under physiological conditions the retinal chromophore brought from pigment epithelium to the rhodopsin molecule, during a regeneration process, in a structure like to planar. From our simulation data we see that the rhodopsin helical distortions helping retinal to be transformed into the nonplanar distorted or twisted conformation. Thus, from our results we suppose that the initial 6-s-cis (like free retinal) transformation requires some initial "re-stabilization" time (say, [0.3-0.4] ns) to occur. During this time interval the rhodopsin helical fluctuations push the retinal to be re-oriented in a manner as described above. So, it is not excluded that the existence of two states of rhodopsin molecule (characterized by two different conformation states of the retinal chromophore) might be possible. These two different forms of rhodopsin molecule may possess different spectral properties and behaviors after the rhodopsin excitation.

Our simulation data described above are in a good agreement with our experimental studies on low-temperature spectroscopy of rhodopsin. Low-temperature spectroscopy results allow to presume that there are more than one intermediate at each stage of rhodopsin photolysis. It is very possible that the rhodopsin photolysis is not a linear process. Probably, there are parallel ways. The product with maximum of absorption at 440 nm is similar to the product known as a blue-shifted intermediate. It has been found in [1] that BSI is the intermediate, which is formed from bathorhodopsin and transformed to lumirhodopsin. In our case, it is clearly to see that this product is not formed from bathorhodopsin, but formed from the product that is appeared simultaneously with classical bathorhodopsin. It can not be excluded that we have observed not one but two products on the early stage of rhodopsin photolysis that is produced from the two forms of chromophore conformation. In other words, one can proposed that there are multiple conformation states of the native dark rhodopsin. This approach is based on our simulation data and other data, concerning the conformation of the beta-ionone ring relative to the chromophore.

Note

From our point of view the theoretical and experimental data, concerning the existence of 6-s-trans-11-cis-retinal configuration in rhodopsin [13, 14, 18, 22], have to be considered as the subject for a more detailed studies. On the one hand, it was shown that the planar 6-s-trans conformation of the free chromophore is energetically favored over the 6-s-cis conformation [25]. On the other hand, the binding site is not tight sufficiently [22], so this fact does not provide enough argument to conclude which conformation is favorable. We have also to remember that the retinal binding pocket is large and enough flexible to accommodate a variety of chromophores [36, 43]. Thus, it is very likely that there is some possibility for the retinal rearrangement (say, from 6-s-cis to 6-s-trans configuration) or for the retinal embedding to be located properly in the opsin site in vivo. We consider the problems discussed above in today are open. Finally, considering the rhodopsin molecule as a template for other G-protein-coupled-receptors (GPCRs) one could suppose that the retinal rearrangement inside the protein binding pocket might be a common process. That is the protein surrounding for other GPRCs may possess the same feature as rhodopsin to adjust for a related ligand or constitutive chain to take (like as chromophore retinal) the most favorable conformation.

References

[1] S.J. Hug, J.W. Lewis, C.M. Einterz, T.E. Thorgeirsson, D.S. Kliger, *Biochemistry*, (1990), 29(6), 1475-1486.

[2] H. Kandori, Y. Shichida, and T. Yoshizawa, *Biochemistry (Mosc)*, (2001), Nov, 66(11), 1197-209.

[3] T. Yoshizawa, G. Wald, *Nature*, (1963), 197, 1279-1285.

[4] L.A. Peteanu, R.W. Schoenlein, Q. Wang, R.A. Mathies, and C.V. Shank, *In: Proc. Natl. Acad. Sci. USA.*, (1993), 90(24), 1176211766.

[5] G. Haran, E.A. Morlino, J. Matthes, R.H. Callender, and R.M. Hochstrasser, *J. Phys. Chem. A*, (1999), 103, 2202-2207.

[6] H. Kandori, Y. Furutani, S. Nishimura, Y. Shichida, H. Chosrowjan, Y. Shibata, N. Mataga, *Chem. Phys. Lett.*, (2001), 334, 271-276.

[7] K. Palczewski, T. Kumasaka, T. Hori, C.A. Behnke, H. Motoshima, B.A. Fox, I. Le Trong, D.C. Teller, T. Okada, R.E. Stenkamp, M. Yamamoto, and M. Miyano, *Science*, (2000), 289, 739-745.

[8] K.-H. Grellmann, R. Livingston, D. Pratt, *Nature*, (1962), 193, 1258-1260.

[9] N. Sasaki, F. Tokunaga, T. Yoshizawa, *Photochem. Photobiol.*, (1980), 32, 433-441.

[10] C.M. Einterz, J.W. Lewis, D.S. Kliger, *In: Proc. Natl. Acad. Sci. USA.*, (1987), 84, 3699-3703.

[11] T.B. Protasova, I.B. Fedorovich, M.A. Ostrovsky, *Light in biology and medicine*, (1991), 2, 545-556.

[12] T.B Fel'dman, I.B Fedorovich, M.A. Ostrovskii, *Neurosci Behav Physiol.*, (2004), Sep, 34(7), 735-42.

[13] P.J.R. Spooner, J.M. Sharples, M.A. Verhoeven, J. Lugtenburg, C. Glaubitz, and A. Watts, *Biochemistry*, (2002), 41, 7549-7555.

[14] M. Sugihara, V. Buss, P. Entel, M. Elstner, and T. Frauenheim, *Biochemistry*, (2002), 41, 15259-15266.

[15] H.G. Smith, G.W. Stubb, B.J. Litman, *Exp. Eye Res.*, (1975), 20, 211-217.

[16] V.A. Krongauz, R.R Shifrina, I.B. Fedorovich, and M.A. Ostrovskii, *Biofizika*, (1975), 20(2): 219-224 (*in Russian*).

[17] T. Yoshizawa, Y. Shichida, S. Matuoka, *Vision Res.*, (1984), 24(11), 1455-1463.

[18] G. Grubner, I.J. Burnett, C. Glaubitz, G. Choi, A.J. Mason, and A. Watts, *Nature*, (2000), 405, 810-813.

[19] S.W. Lin, M. Groesbeek, I. van der Hoef, P. Verdegem, J. Lugtenburg, and R.A. Mathies, *J. Phys. Chem. B*, (1998), 102, 2787-2806.

[20] P.J.E. Verdegem, P.H.M. Bovee-Geurts, W.J. de Grip, J. Lugtenburg, and H.J.M. De-Groot, *Biochemistry*, (1999), 38, 11316-11324.

[21] M. Han and S.O. Smith, *Biochemistry*, (1995), 34, 1425-1432.

[22] D. Singh, B.S. Hudson, C. Middleton, and R. Birge, *Biochemistry*, (2001), 40(14), 4201-4204.

[23] T. Okada, M. Sugihara, A.-N. Bondar, M. Elstner, P. Entel and V. Buss, *J. Mol. Biol.*, (2004), 1-13.

[24] A.B. Patel, E. Crocker, M. Eilers, A. Hirshfeld, M. Sheves, and S.O. Smith, *PNAS*, (2004), 101(27), 10048-10053.

[25] F. Terstegen, and V. Buss, *J. Mol. Struct. (THEOCHEM)*, (1998), 430, 209-218.

[26] I.D. Pogozheva V.A. Kuznetsov, V.A. Livshits, I.B. Fedorovich, M.A. Ostrovsky, *Biologicheskie Membrany (Biological Membranes)*, (1985), 2, 880-896 (in russian).

[27] I.D. Pogozheva, V.A. Kuznetsov, V.A. Livshits, M.A. Ostrovsky, *Biologicheskie Membrany (Biological Membranes)*, (1985), 2, 897-905 (in russian).

[28] S.J. Weiner, P.A. Kollman, D.T. Nguyen, D.A. Case, *J. Comp. Chem.*, (1986), 7, 230.

[29] B.R. Brooks, R.E. Bruccoleri, B.D. Olafson, S. States, D.J. Swanimathan, M. Karplus, *J. Comput. Chem.*, (1983), 4, 187-217.

[30] T. Okada et al., *J. Struct. Biol.*, (2000), 130, 73-78.

[31] T. Shieh, M. Han, T.P. Sakmar, S.O. Smith, *J. Mol. Biol.*, (1997), 269, 373-377.

[32] P. Herzyk, R.E. Hubbard, *J. Mol. Biol.*, (1998), 281, 741-747.

[33] R. Henderson, G.F.X. Schertler, *Phis. Trans. R. Soc. Lond.*, (1990), 326, 379-389.

[34] P.S. Crozier, M.J. Stevens, L.R. Forrest, T.B. Woolf, *J. Mol. Biol.*, (2003), 333, 493-514.

[35] E. Zrenner, *Science*, (2002), 295, 1022-1025.

[36] K. Palczewski, T. Kumasaka, T. Hori, C.A. Behnke, H. Motoshima, B.A. Fox, I. le Trong, D.C. Teller, T. Okada, R.E. Stenkamp, M. Yamamoto, M. Miyano, *Science*, (2000), 289, 739-745.

[37] D.C. Teller, T. Okada, C.A. Behnke, K. Palczewski, R.E. Stenkamp, *Biochem.*, (2001), 40, 7761-7772.

[38] K. Kandori, I. Shichida, T. Yoshizawa, *Biokhimiya*, 2001, 66, 1483-1498 (*in Russian*).

[39] D. Pan, R.A. Mathies, *Biochem.*, (2001), 40, 7929-7936.

[40] W. Humphrey, H. Lu, I. Logunov, H.J. Werner, K. Schulten, *Biophys. J.*, (1998), 75, 1689-1699.

[41] E. Tajkhorshid, J. Baudry, K. Schulten, S. Suhai, *Biophys. J.*, (2000), 78, 683-693.

[42] S. Hayashi, E. Tajkhorshid, K. Schulten, *Biophys. J.*, (2002), 83, 1281-1297.

[43] J. Saam, E. Tajkhorshid, S. Hayashi, K. Schulten, *Biophys. J.*, (2002), 83, 3097-3112.

[44] A. Warshel, *Nature*, (1976), 260, 679-683.

[45] R.R. Birge, L.M. Hubbard, *J. Am. Chem. Soc.*, (1980), 102, 2195-2205.

[46] A. Warshel, N. Barboy, *J. Am. Chem. Soc.*, (1980), 104, 1469-1476.

[47]. J.R. Tallent, E.W. Hyde, L.A. Findsen, G.C. Fox, R.R. Birge, *J. Am. Chem. Soc.*, (1992), 114, 1581-1592.

[48] U. Rohrig, L. Guidoni, U. Rotlisberger, *Biophys. J.*, (2002), 82, 223a.

[49] G. Choi, J. Landin, J.F. Galan, R.R. Birge, *Biochem.*, (2002), 41, 7318-7324.

[50] K. Kholmurodov, T. Ebisuzaki, *ICMS-CSW2004*, Tsukuba, (2004), C4, 9-11.

[51] K. Kholmurodov, et al., *PEPAN (Physics of Elementary Particles and Atomic Nuclei)*, 34(2), 474-510, JINR, Dubna, Moscow, 2003.

[52] T.P. Sakmar, R.R. Franke, H.G. Khorana, *Prot. Nat. Acad. Sci. USA*, (1989), 86, 8309-8313.

[53] R.R. Birge, *Biochim. Biophys. Acta.*, 1990, 1016, 293-327.

[54] D.L. Farrens, C. Altenbach, K. Yang, W.L. Hobbell, H.G. Khorana, *Science*, (1996), 274, 768-780.

[55] H.G. Khorana, *Prot. Nat. Acad. Sci. USA.*, (1993), 90, 1166-1171.

[56] U. Gelther, B.K. Kobilka, *J. Biol. Chem.*, 1998, 273, 17979-17982.

[57] O. Kuwata, C. Yuan, S. Misra, R. Govinjee, T. Ebrey, *Biochemistry*, (2001), 66, 1588-1608 (in Russian).

[58] S.P. Sheikh, T.A. Zvyaga, O. Lichtarge, T.P. Sakmar, H.R. Bourne, *Nature*, (1996), 383, 347-350.

[59] Z. Farahbakhsh, K. Hideg, W. Hubbell, *Science*, (1993), 262, 1416-1419.

[60] W. Hubbell, D. Cafiso, C. Altenbach, *Nature Stuct. Biol.*, (2000), 7, 735-739.

[61] C. Altenbach, J. Klein-Saathezaraman, K. Cai, H.G. Khorana, W.I. Hubbell, *Biochemistry*, 2001, 40, 15493-15500.

[62] *MOE (Molecular Operating Environment)*, used within 2002-2003, by license of CAL RIKEN, .

[63] D.A. Case, D.A. Pearlman, J.W. Caldwell, T.E. Cheatham III, W.S. Ross, C.L. Simmerling, T.A. Darden, K.M. Merz, R.V. Stanton, A.L. Cheng, J.J. Vincent, M. Crowley, D.M. Ferguson, R.J. Radmer, G.L. Seibel, U.C. Singh, P.K. Weiner, P.A. Kollman, *AMBER 5*, University of California, 1997.

[64] T. Narumi, R. Susukita, T. Ebisuzaki, G. McNiven, B. Elmegreen, *Molecular Simulation*, (1999), 21, 401-408.

[65] T. Narumi, R. Susukita, H. Furusawa, T. Ebisuzaki, *IN: Proc. 5th Int. Conf. on Signal Processing*, (2000), Beijing, 575-582.

[66] N. Okimoto, K. Yamanaka, A. Suenaga, Y. Hirano, N. Futatsugi, T. Narumi, K. Yasuoka, R. Susukita, T. Koishi, H. Furusawa, A. Kawai, M. Hata, T. Hoshino, T. Ebisuzaki, *Chem-Bio Informatics J.*, (2003), 3(1), 1-11.

[67] W.D. Cornell, P. Cieplak, C.I. Bayly, I.R. Gould, K.M. Merz Jr, D.M. Ferguson, D.C. Spellmeyer, T. Fox, J.W. Caldwell, P.A. Kollman, *J. Am. Chem. Soc.*, (1995), 117, 5179-5197.

[68] W.L. Jorgensen, J. Chandrasekhar, J.D. Madura, *J. Chem. Phys.*, (1983), 79, 926-935.

[69] H.J.C. Berendsen, J.P.M. Postma, W.F. van Gunsteren, A. DiNola, J.R. Haak, *J. Chem. Phys.*, (1984), 81, 3684-3690.

[70] J.P. Ryckaert, G. Ciccotti, H.J.C. Berendsen, *J. Comput. Phys.*, (1997), 23, 327-341.

[71] R.A. Sayle, E.J. Milner-White, RasMol: Biomolecular graphics for all, *Trends in Biochem. Sci.*, (1995), 20, 374-376.

[72] R. Koradi, M. Billeter, K. Wüthrich, MOLMOL: a program for display and analysis of macromolecular structure, *J. Mol. Graphics*, 1996, 4, 51-55.

[73] J. Kim, C. Altenbach, R. Thurmond, H.G. Khorana, W.I. Hubbell, *Prot. Natl. Acad. Sci. USA.*, (1997), 94, 14273-14278.

[74] R. Franke, B. Konig, T. Sakmar, H. Khorana, *Science*, (1990), 250, 123-125.

[75] Fahmy et al., *In: Proc. Natl. Acad. Sci. USA*, (1993), 90, 10206-10210.

In: Molecular Simulation Studies in Material ... ISBN 1-59454-607-X
Editor: Kholmirzo Kholmurodov, pp. 115-128

Chapter 9

STUDY OF CLUSTER STATE OF FULLERENES IN SOLUTIONS: CURRENT STATUS AND PROSPECTS

M. V. Avdeev, T. V. Tropin and V. L. Aksenov*
Frank Laboratory of Neutron Physics, Joint Institute for Nuclear Research
141980, Dubna, Moscow Reg., Russia

Abstract

Experimental and theoretical aspects of the study of fullerene clusters in molecular and colloidal solutions are presented. Experimental data obtained by different techniques including small-angle neutron scattering from a number of solutions of fullerenes are summarized covering molecular solutions of C_{60} fullerene in carbon disulfide, colloidal solutions of C_{60} fullerene in water, nitrogenated solvents and binary liquids. Conditions and mechanisms of cluster formation in these solutions are considered. Applications of molecular dynamics simulations for modeling the considered solutions are discussed.

1. Introduction

Fullerenes, recently discovered new allotropic form of carbon (C_{60}, C_{70} and others), are of current interest. In contrast to other forms of carbon (graphite and diamond), fullerenes are soluble [1], [2] in a number of solvents including non-polar organic solvents (maximal solubility 50 mg/ml of C_{60} in chlornaphtaline), as well as carbon disulfide, nitrogen-containing solvents (e.g. pyridin, 1-methyl-2pyrrolidon) and others. However, the solubility of fullerenes in polar solvents (e.g. alcohols) is highly low [1], [2]. Thus, fullerenes are insoluble in water.

Three basic points of interest are connected with the fullerene solutions. First, the knowledge about their properties may be used in production of fullerenes, in particular when separating different types of fullerenes in mixtures [3], [4]. Second, fullerenes exhibit biological activity and have much prospects for medical applications. For example, the reported antioxidant properties can be employed in cancer treatment [5], [6] and curing of neurological diseases [7], [8]. Other applications. Inhibition activity with respect

*E-mail address: avd@nf.jinr.ru

to proteases in HIV [9], as well as to grippe viruses [10] should be mentioned. Since true solutions of fullerenes in polar biological media are impossible, the effective use of fullerenes in medicine depends on production of stable and non-toxic colloidal solutions of fullerenes and their derivatives in polar solvents, particularly in water [11]-[14]. The third point concerns the problem of cluster state of fullerenes in different solutions. The large size of fullerene molecules determines the unusual behavior of fullerene solutions. They are characterized by a complex kinetics of dissolution, which can endure for months [15]. This feature is tightly connected with the tendency for fullerenes in solutions to form clusters. The cluster formation of fullerenes was reported [16]-[21] in different types of solvents. It was assumed [1], [22] to be a reason for anomalous temperature dependence of solubility discovered in a number of solvents, as well as solvatochromatic effects in polar liquids and their mixtures. However, often the experimental data are contradictory, so the complex and detailed study of this phenomenon is required.

In the given paper we review the current state in the study of fullerene clusterization both in molecular and colloidal solutions. Experimental and theoretical aspects are presented. The following directions are covered. They are the formation of fullerene clusters in molecular solutions (Chapter 2.), aqueous dispersions of fullerenes (Chapter 3.), aggregation in nitrogen-containing solvents and their mixtures with water (Chapter 4.). A special attention is drawn to the possibility to use molecular dynamics simulations in study of fullerene solutions.

2. Clusterization of Fullerenes in Molecular Solutions

The problem of cluster formation in molecular solutions of fullerenes is connected with the anomalous temperature dependence of their solubility, $c_{max}(T)$. First, it was discovered by Ruoff, et al. in 1993 for solutions of C_{60} in non-polar solvents including toluene, hexane and carbon disulfide [23]. Later, the similar effect was observed in xylene [24]. In the indicated solvents at low temperatures the solubility of C_{60} grows with temperature like in most solutions. However, after it achieves the maximal value at T_m ~280 K, it decreases. The interesting feature of this effect was that the maximum of solubility takes place approximately at one temperature despite the significant difference (up to two orders of magnitude) in absolute value of solubility for these solutions. This fact, as well as the closeness of T_m to the critical temperature corresponding to the phase transition of the first order in fullerene crystals of C_{60} ($T_c = 260$ K) allowed one to connect the specific behavior of the solubility with this phase transition. The idea was that for the crystal phase, which should be in the equilibrium with the saturated solution, at $T < T_c$ the enthalpy of dissolution is positive, while for the phase at $T > T_c$, is negative. The later results in a decrease of solubility with the temperature growth. It is clear that the difference of these enthalpies should be equal to the enthalpy of the phase transition, which is, however, not true for the studied systems. It is of about five times more, than the enthalpy of the phase transition obtained experimentally. To explain this significant disagreement it was assumed [1], [22] that, in addition to the single state, fullerenes in solutions are present in the form of thermodynamically stable clusters, so that the equilibrium concentration of fullerenes (solubility) is determined by the equilibrium between solid phase, single fullerenes in solvent and fullerenes in clusters. The reason for this suggestion was the direct observation of fullerene clusters in a number of organic non-polar solvents, such as benzene [16], [17], toluene [18], carbon disulphide [19] and others. Using the liquid droplet model to describe the cluster formation, the cluster size distribution function was determined [1], [22] from conditions of thermodynamic equilibrium between three phases. It is different significantly for the temperatures below and higher than T_c. At $T < T_c$ the role of clusters is minimal, the fullerene is mainly in the form of free monomers. At $T > T_c$ the clusters start affect much the temperature dependence of solubility, the main part of dissolved fullerenes is in clusters, which determines its anomalous dependence. While the given phenomenological theory looks reasonable, there are contradictory works showing the evidence that fullerenes are present as single molecules in non-polar solvents (e.g. [25]-[28]). That is why, an alternative approach [29]-[31] was developed. The explanation of the discussed effect was given basing on the idea of formation and melting of crystal solvates. At temperatures below T_m the saturated solution is not in the equilibrium with the crystal of fullerenes, but with solvated crystal, when the solvent molecules, due to donor-acceptor bonds, are included in crystal structure. At $T > T_m$ the crystal solvate melts, and the pure fullerene crystal is in equilibrium with the solution. The enthalpy of dissolution of fullerene crystals should be negative in this case, which is testified experimentally. It should be pointed out that in the frame of this model the T_m temperature is not connected with the phase transition temperature in the C_{60} crystal. One can say two facts in favor of this model. First, there is a solvent (brome-benzene) where the peak of solubility takes place at $T_m = 350$ K, which is different much from that indicated for other solvents. The other one is the temperature behavior of the C_{70} fullerene solubility. According to [1], [22], [23] there should be no anomaly in the $c_{max}(T)$ function in this case, since no phase transition in crystals of C_{70} is observed at room temperatures. Actually, it was confirmed in a number of C_{70}, where a monotonous increase in the solubility with the temperature was observed. However, recently the anomalous temperature dependence of the solubility was reported [24], [32] for solutions of C_{70} in 1,2-dichlorbenzene ($T_m = 328$ K) and 1,2-dimethylbenzene ($T_m = 283$ K).

Thus, the clarification of the role of fullerene clusters discovered in non-polar solutions in the anomalous behavior of the temperature is required. For this purpose, different experimental methods and theoretical approaches can be used. Among the experimental techniques are electron microscopy (EM), positron annihilation spectroscopy (PAS), dynamic light scattering (DLS), small-angle neutron scattering (SANS). These techniques make it possible to detect clusters in solutions and describe the cluster size distribution function. We concern here some details of the study of this problem obtained recently for fullerene solutions in carbon disulfide (CS_2) with a comparatively high solubility, $c_{max} = 7.9$ mg/ml [1], [2] at room temperature. First applications of the small-angle neutron scattering for these solutions showed [27] that despite the small size of fullerene molecules they produce observable scattering signal which can be treated in the standard way. No clusters were reported. Nevertheless, the visible size of fullerene in these experiments was about 10 % higher than that calculated from atomic coordinates of C-atoms in the fullerene molecule. It was supposed [27], [28], [33] that the origin of this effect is connected with the systematic error of the Guinier approximation used to estimated the size. Later, such experiments were repeated several times [28], [33]-[36]. Different methods of data treatment including direct modeling and indirect Fourier transform which are of better precision than the Guinier approximation also resulted in the indicated difference. Several reasons for this effect can be

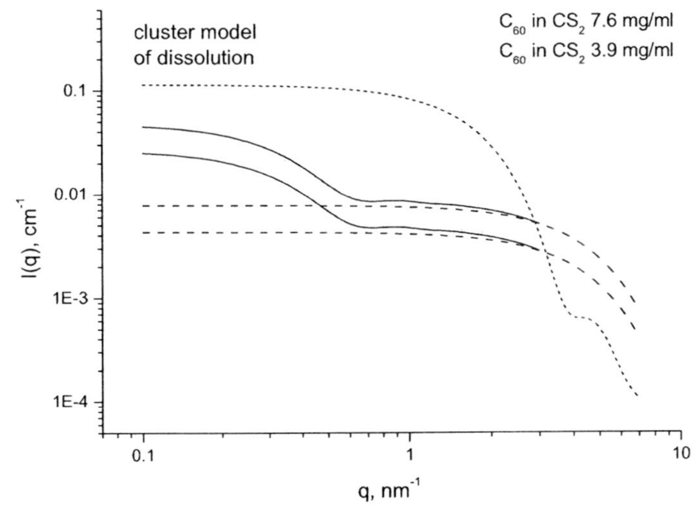

Figure 1. Experimental SANS curves (points) from solutions of C_{60} fullerene in CS_2 at different concentrations (obtained at the IBR-2 pulsed reactor, JINR, Dubna). Deviation at small q-values from the model scattering of single fullerenes (dashed lines) is observed. Solid lines correspond to the model taking into account the clusters in the solutions. The resulting function of the cluster size distribution is given in Fig.2

discussed. The first one is the formation of the solvation shell around fullerenes due to their interaction with solvent molecules [34]. This assumption was checked carefully in [33] by measuring the concentration dependence of the scattering intensity and estimating the second virial coefficient of the solution. The conclusion was that this coefficient is close to zero, so there is no significant interaction between fullerenes and solvent molecules which would result in the solvation shell. Another reason is connected with the possible small aggregation, which is also suggested [26] for some fullerene solutions. If, in addition to monomers, there are dimers, trimers, etc. in the solution and their fraction is not high, the character of the scattering curve is similar to that from the solution of monomers but with different parameters (in particular, the characteristic particle size) affected by the presence of aggregates. Finally, the discussed difference between the calculated and visible size of fullerenes in CS_2, can be caused by the so-called unacceptable volume effect. Usually, to calculate the volume of the dissolved molecule in a solvent the molecule of the latter is approximated by a sphere. From the covering of the dissolved molecules by such spheres the acceptable surface and volume unacceptable to solvent are estimated (e.g. [37], [38]. It is clear that for the good precision the size of the solvent molecule should be significantly smaller in comparison with that of the dissolved molecule. This is not the case for fullerene solutions in CS_2, so that a specific organization of the CS_2 molecules close to fullerene surface would result in an effective increase of the unacceptable volume, and, hence, in visible size, of the fullerene. This assumption may be checked well by the molecular dynamics simulations. The results of such modeling is presented below in these Proceedings [39]. The main conclusion of this work is that there is no effective increase in the size of the fullerene molecule, but a modulation of the solvent density on its surface. It is similar to a short ordering because of the intermolecular interaction in liquids, which determines the characteristic view of the radial distribution function, $g(r)$. This means that the situation is close to the formation of the solvation shell. The disagreement of this conclusion with the SANS analysis [33] can be explained by the small volume fraction of the dissolved fullerene ($<$ 0.6 %), so that it is hardly possible to observe the interaction effect in SANS data. Moreover, large fullerene clusters are possible in this solution, which decreases the free surface of the fullerene, and, hence, makes the interaction effect less significant. These clusters were observed in recent experiments on the small-angle neutron scattering and in the work on the transmission electron microscopy and positron annihilation spectroscopy [19]. The clusters differ much from those predicted in the cluster model of dissolution [1], [22]. Their structural features can be judged from the SANS curves shown in Fig.1. These curves differ from the ones measured previously by the smaller q-values covered. This q-region allowed us to detect the scattering deviating from that of single fullerene (Fig.1) and conclude about the presence of quite large clusters in the solutions. The model curves in Fig.1 correspond to the particle size distribution shown in Fig.2. The size of the fullerene clusters is about 8 nm, while the fraction of fullerenes in clusters is not high, $\sim$0.005. In this experiment the particle size distribution does not change with the concentration, which is clearly seen in Fig.1 as the similarity of the data obtained for two concentrations. However, it was discovered that it depends strongly on the preparation procedure, namely on the non-equilibrium effect of ultrasound treatment, stirring and their duration when dissolving fullerene. According to results of the EM and PAS experiments [19], the cluster structure in the C_{60}/CS_2 system is concentration dependent. Several regimes of the cluster growth can be distinguished at low concentration below 2.15 mg/ml. The thorough study of this effect is in progress.

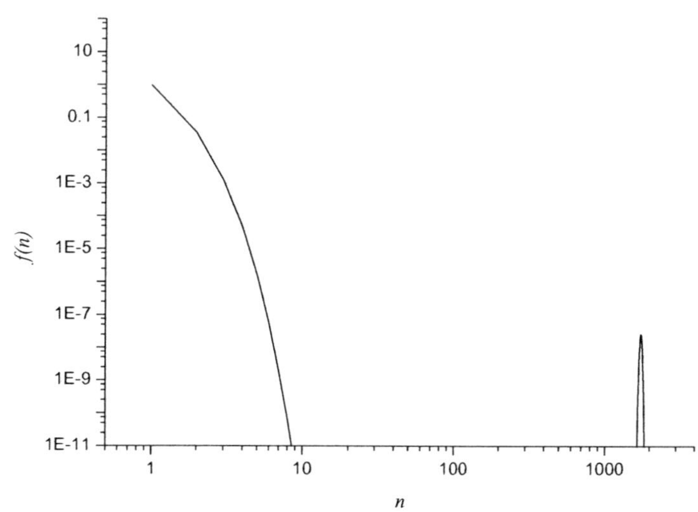

Figure 2. Distribution of the cluster size n (number of monomers in cluster) as revealed by SANS. The solution is prepared with the help of ultrasound

Recently, the possibility of the cluster formation in C_{60}/CS_2 was considered theoretically using methods of the nucleation theory [40]. The result of the time evolution of the cluster size distribution function in the frame of this theory is the formation of the infinite cluster reflecting the appearance of the new phase. Nevertheless, the transform to the infinite cluster may take large periods depending on the parameters of the system, as well as mechanisms of clusterizaton. In this case the growing clusters change very slowly and correspond to a metastable state. With respect to the discussed problem the task was to estimate the characteristic time of this transform for fullerene solutions based on experimentally found diffusion coefficient of the solutions. First, it was done for the liquid drop model of clusters, when the expression for the work of formation of the cluster, $\Delta G(n)$, comprising n monomers combines two terms corresponding to the volume and surface energy, respectively:

$$\Delta G(n) = -n\Delta\mu + \alpha_2 n^{2/3}. \tag{1}$$

Here, $\Delta\mu$ is the difference between chemical potentials of the fullerene molecule in free state and in the cluster. It is determined by the fullerene concentration, c, as following:

$$\Delta\mu = k_B T \ln\left(\frac{c}{c_{max}}\right). \tag{2}$$

The α_2 coefficient characterizes the surface tension, σ, at the interface between fullerene and solvent:

$$\alpha_2 = 4\pi\sigma(3V_0/4\pi)^{2/3}, \tag{3}$$

where V_0 is the fullerene volume unacceptable for the solvent. It should be pointed out that in the frame of this model clusters can appear only if the initial concentration of fullerenes put into the solutions is more than the equilibrium concentration c_{max}. This corresponds to metastable initial conditions and reflects well non-equilibrium ways of the solution preparation which affect the cluster formation as it was mentioned above. From the numerical solution of kinetics equations of the nucleation theory [41] the time evolution of the cluster size distribution function was obtained and analyzed. Found characteristic times of the infinite cluster growth for fullerene solutions were less than 1 ms, which means that the liquid drop model for the clusters does not satisfy the experimental data. The natural development of this model is the suggested model of the limited cluster growth. In contrast to the previous case, an additional size-stabilizing term is added in Eq.3:

$$\Delta G(n) = -n\Delta\mu + \alpha_2 n^{2/3} + kn^{\beta}, \tag{4}$$

where k and β are parameters responsible for the limitation of the cluster size. The principle differencé between two equations is demonstrated in Fig.3. In this case the cluster size distribution function approaches fast to the type given in Fig.4, where the peak position is determined by the size corresponding to the minimum in the $G(n)$ function. The model is appropriate for describing obtained experimental data. However, the origin of the size limitation is to be clarified.

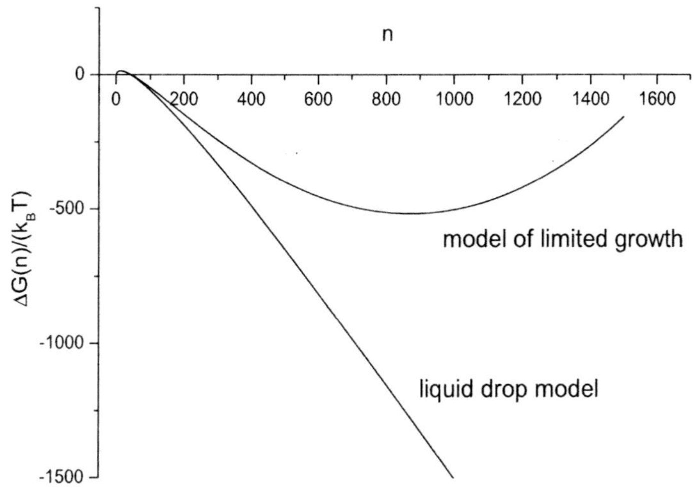

Figure 3. Work of cluster formation as function of the cluster size, *n*, for two models (see text) at typical values of molecular fullerene solutions

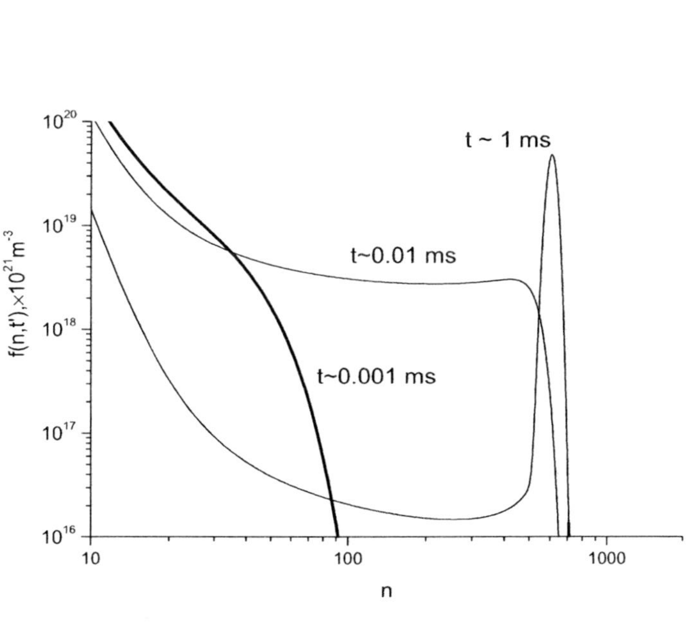

Figure 4. Evolution of the cluster size distribution function in the system C_{60}/CS_2 calculated for different orders of time according to the model of limited growth

From this Chapter one can conclude that molecular solutions of fullerenes in non-polar solvents are often characterized by the cluster formation. The appearance of clusters cannot be explained in the frame of a simple liquid drop model. The model of the limited cluster growth is the appropriate development of the liquid drop model. It describes phenomenalogically a stabilization mechanism, while the origin of this mechanism is unclear at the moment. Molecular dynamics simulations reveal a specific organization of the solvent molecules around fullerene molecules, whose effect on the stabilization of fullerene clusters should be studied. The influence of the non-equilibrium dissolution on the cluster formation has been detected. The problem of the role of fullerene clusters in the anomalous behavior of solubility in molecular solutions of fullerenes is open. Nevertheless, it should be pointed out that there is no direct evidence that fullerene clusters determine such behavior.

3. Aqueous Fullerene Dispersions

As it was mentioned in Introduction, fullerenes cannot form true solutions in water. Along with it, there are several ways to produce aqueous colloidal solutions of fullerenes. The first way is the solubilization, i. e. the use of polar surfactant micelles containing fullerenes. Examples of such solutions stabilized with polyvinylpyrrolidon and Triton X-100 can be found in [42]. The second way is the design of fullerene derivatives soluble in water. For example, the solubility in water is achieved [43] for the compound C_{60}-$((CH_2)_5SO_3Na)_6$, the star-like complex with hydrophilic tails attached to the fullerene. Its structure determines a specific aggregation in aqueous solutions. Finally, aqueous fullerene dispersions can be produced directly by the solvent substitution method recently developed [11]-[14]. The appearance of such dispersions gives promise that the biological activity of unmodified fullerenes may be used. The study of these dispersions obtained by this method will be discussed below.

The method does not require any stabilizer or chemical modification of fullerene. It is based on exchange of molecules of an organic solvent, which cover fullerenes, with water molecules. There are a number of reported schemes how to do this. Here, we consider in details the dispersions produced by Andrievsky and co-workers [12], [44], [45]. According to the corresponding procedure, a true fullerene solution in an organic solvent (toluene, benzene), which is non-miscible or slightly miscible with water, is prepared. Then, it is dispersed into aqueous phase with the help of ultrasound of high power. During this process the organic solvent is evaporated from the system resulting in formation of stable colloidal aggregates (clusters). Finally, a special chemical treatment to remove remains of the organic solvent is performed. Thorough analysis of the produced dispersions showed that there are no inclusions of the organic solvent into the cluster structure. The stabilization mechanism of such dispersions is not clear for the moment, which raises an interesting problem from the viewpoint of colloidal chemistry. Nevertheless, a negative charge of colloidal particles detected in different works [13], [44] (ζ-potential 9÷30 mV) seems to play a significant role in the stabilization, while its origin is still under discussion.

The structure of aqueous fullerene dispersions is studied by techniques which are usually employed in the case of colloidal solutions. They are the electron microscopy (EM), dynamic light scattering (DLS), small-angle neutron (SANS) and X-ray scattering (SAXS), sedimentation analysis (SA). Also, the differential scanning calorimetry (DSC) gives interesting observations. All of the methods showed the characteristic size close to 70 nm for

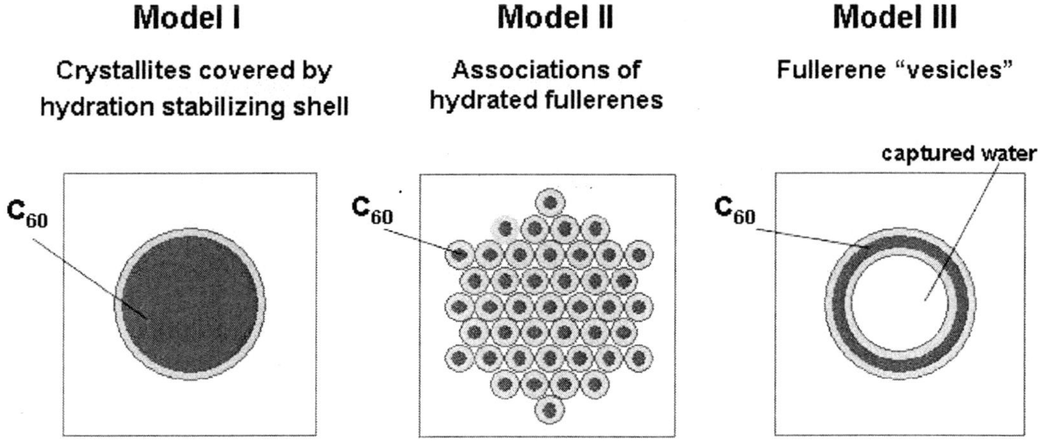

Figure 5. Models of fullerene aggregates in aqueous dispersions

colloidal particles in the system. The dispersions are highly stable (more than 18 months), and their preparation procedure is well reproducible. Taking into account the results of the complex study of the dispersions three possible models are discussed actively [46]. They are presented schematically in Fig.5. Model I in this Figure implies that the particles are fullerene crystallites or rather dense clusters. A strong argument for this model is the co-incidence of the densities of the clusters and fullerene crystals. Probably, the stabilization of these clusters is provided by charged thin shells of hydration water, the result of donor-acceptor bonds between surface fullerene molecules and water. In the frame of this model the fraction of the hydration shell in the particles must not exceed 5 vol. %. Model II (Fig.5) suggests that the particles represent associates of single hydrated C_{60} fullerenes. In this case hydration water must be localized on the fullerene surface, and the maximum number of water molecules taking part in the hydrolysis is equal to six. Model III (Fig.5) is intended to explain an additional peak in the DCS traces different from water peak [47]. It considers the particles in the dispersions as closed film-like structures with water captured inside. The estimated size of such inclusion is of the same order as the particle size. The free exchange of the captured and bulk water must take place. In contrast to dispersions produced by other schemes of the organic solvent substitution, the considered dispersions have rather wide particle size distribution function, which we connect with the effect of ultrasound treatment. This fact determines a principle difficulty in the study of these dispersions. There is no simple reliable way to conclude about the particles with the size less than 10 nm. In all techniques the signals from larger particles prevail. Even, if the size separation is made, the concentration of small particles is rather low to analyze them quantitatively, so a special procedure to increase this concentration should be designed. Nevertheless, some indications of the presence in the system of small particles have been reported basing on the data of chemical tests and small-angle neutron scattering.

Calculations of preferable fullerene cluster configurations in aqueous dispersions were reported [48]. They took into account the interaction of fullerene with water and resulted in icosahedral systems (I_h symmetry group). Such symmetry allows some discrete values for the aggregation number of the stable fullerene clusters from the series 13, 33 and 63.

Interaction of fullerene molecules with water is one of the most important questions which could clarify the stabilization mechanism of the discussed aqueous dispersions of fullerenes. We suppose that applications of MDS in this case can help to solve the problem.

4. Aggregation of Fullerenes in Nitrogenated Solvents

Fullerene solutions based on nitrogen-containing solvents are distinguished as a separate class due to the solvatochromatic effect, i.e. changes in the UV-Vis spectrum under different conditions. At small rate this effect takes place in time in pure solutions and is connected with the formation of clusters [49]. It is significantly stronger in the case of the solvents miscible with water and other polar solvents (methanol, acetonitrile), if the last solvents are added in the solution. This effect was discovered and investigated thoroughly in the system C_{60}/pyridine/water [49]-[51]. The solution of C_{60} in pure pyridine has a mauve color. The added water with the relative content of more than 50 vol. % changes sharply the color to lemon-yellow. The UV-Vis spectrum becomes similar to that from aqueous dispersions of fullerenes considered above. It is accompanied by reorganization in the cluster size distribution. In pure pyridine solutions the clusters are quite large (characteristic size $>$ 100 nm), while in the binary mixture they are very similar to those observed in water dispersions while the characteristic size is less $\sim$ 40 nm. It was supposed [49] that the cluster stabilization in these solutions is based on the formation of a specific pyridine shell around the clusters, which was testified indirectly by chemical tests with hydrazine. For molecular fullerene solutions, in particular for pure pyridine, hydrazine reacts with the fullerene and is a good indicator of its presence. In mixtures with water no reaction with hydrosine takes place, which is an argument for the existing of the shell. However, closeness of the UV-Vis spectra with that from aqueous dispersions points to the fact that water can also play a significant role in stabilization of the considered mixtures. One can see that the addition of water in the system increases significantly the number of smaller particles in comparison with the pure pyridine solutions. At the moment, it is difficult to conclude whether this is a result of the destruction of large clusters (decrease in the characteristic size of large particles) or the development of small clusters (increase in the characteristic size of small particles). To clarify this point further experiments are required. Recently, this effect was confirmed [52] in other system based on nitrogen containing solvent 1-methyl-pyrrolidon (NMP) also miscible with water. The solvatochromism is not so pronounced as in the previous case, the solution does not change its color sharply and only small changes in UV-Vis spectrum take place. But the structural reorganization of the clusters on addition of water is similar. This can be judged from Fig.6, where the small-angle scattering from different solutions are presented. Despite a small SANS signal in pure NMP one can conclude that quite large clusters (size more than 100 nm) were present in the solutions. This is testified by the fact that the signal decreases with the decrease in the fullerene concentration. This was also proved by complementary data on UV-Vis absorption and dynamic light scattering (DLS). When water is added in the system the SANS signal increases significantly, which is a reflection of appearance of the clusters with characteristic size of about 25 nm in the mixture. It is interesting that if the NMP solution is mixed with deuterated water such SANS signal does not take place. This means that the mean scattering length densities of the clusters and the solvent in this case are close, which, in turn, points to comparatively small NMP content in the clusters. Qualitatively, the situation repeats that with the pyridine solutions.

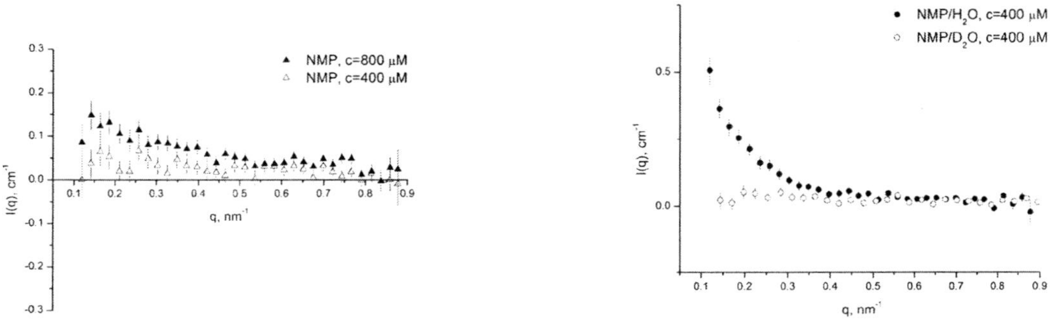

Figure 6. Small-angle neutron scattering from NMP (*a*) and NMP/water (*b*) solutions of C60 fullerene

It should be noted that mixtures of nitrogen-containing solvents with water are of a special interest themselves. Specific correlation between molecules is observed in these systems, which is a subject for molecular dynamics simulations [53]. Often these systems are characterized by effects of phase separation [54].

Thus, the nitrogen-containing solvents show similar effects both in pure solutions and mixtures with water. On the one hand one can connect this with specific chemical interactions of these solvents with fullerenes. On the other hand, some common features with respect to the aqueous fullerene dispersions can be distinguished, which makes the study of these systems as of special interest.

5. Conclusions

One can see that in all presented directions of the study of cluster state of fullerene in solutions common features can be found. This means that the development of these directions simultaneously with the help of experimental, theoretical and modeling approaches will help to clarify the problem of the fullerene cluster formation. The more specificity of the fullerene clusters should be mentioned in conclusion. It is connected with the fact that the structural unit of fullerene clusters, which are fullerene molecules, is of known structure and monodisperse. This is quite seldom case in the cluster growth processes taking place in nature. This makes fullerene solutions unique model systems where different aspects of the theory of the cluster state can be effectively studied.

Acknowledgments

The work has been performed with the support of the Russian Ministry Education and Science, state contract №40.012.1.1.1148.

References

[1] V.N. Bezmelnitsyn, A.B. Eletskii, M.B. Okun, *Usp. Fiz. Nauk*, (1998), 168, 1195 (*in Russian*).

[2] M.V. Korobov, A.L. Smith, *Fullerenes: chemistry, physics and technology*, K.M. Kadish, R.S. Ruoff eds., A John Wiley&Sons, Inc., 2000, 53-89.

[3] W. Kraetschmer, et al., *Nature (London),* (1990), 347, 354.

[4] R. Taylor, et al., *J. Chem. Soc. Chem. Commun.*, (1990), 1423.

[5] A.D. Roslyakov, et.al., *Zh. Akad. Med. Nauk Ukrainy*, (1999), 338, 5 (*In Russian*).

[6] T. Da Ros, M. Prato, *Chem. Commun.*, (1999), 663.

[7] L.L. Dugan, J.K. Gabrielsen, Sh.P. Yu, T.-S. Lin, D.W. Choi, *Neurobiology of Decease*, (1996), 3, 129.

[8] I.Ya. Podolsky, E.V. Kondratiev, I.V. Scheglov, M.A. Dumpis, L.B. Piotrovsky, *Physics Solid State*, (2002), 44 (3), 552.

[9] R. Sijbesma, G. Srdanov, F. Wudl, J.A. Castro, et al., *J. Am. Chem. Soc.*, (1993), 115, 6510.

[10] L.B. Piotrovsky, et al., *Mol. Mater.*, (2000), 13, 41.

[11] W.A. Scrivence, J.M. Tour, *J. Am. Chem. Soc.*, (1994), 116, 4517-4518.

[12] G.V. Andrievsky, M.V. Kosevich, O.M. Vovk, V.S. Shelkovsky, L.A. Vashchenko, *J. Chem. Soc., Chem. Commun.*, (1995), 12, 1281-1282.

[13] S. Deguchi, R.G. Alargova, K. Tsujii, *Langmuir*, (2001), 17, 6013-6017.

[14] P. Scharff, K. Risch, L. Carta-Abelmann, I.M. Dmytruk, M.M. Bilyi, O.A. Golub, A.V. Khavryuchenko, E.V. Buzaneva, V.L. Aksenov, M.V. Avdeev, Yu.I. Prylutskyy, S.S. Durov, *Carbon*, (2004), 42, 1203-1206.

[15] T. Tomiyami, S. Uchiyama, H. Shinohara, *Chem. Phys. Lett.*, (1997), 264, 143-148.

[16] Q. Ying, J. Marecek, B. Chu, *Chem. Phys. Lett.*, (1994), 219, 214-218.

[17] Q. Ying, J. Marecek, B. Chu, *J. Chem. Phys.*, (1994), 101, 2665.

[18] Gy. Torok, V.T. Lebedev, L. Cser, *FTT*, (2002), 44(3), 546 (*In Russian*).

[19] A.D. Bokare, A. Patnaik, *J. Chem. Phys*, (2003), 119, 4529.

[20] S. Nath, H. Pal, D.K. Palit, A.V. Sapre, J.P. Mitttal, *J. Phys. Chem. B.*, (1998), 102, 10158.

[21] J.S. Ahn, K. Suzuki, Y. Iwasa, N. Otsuka, T. Mitani, *Proc. SPIE Fullerenes and Photonics*, IV, SPIE, Bellingham, WA, 1997, 3142.

[22] V.N. Bezmelnitsin, A.V. Eletskii, E.V. Stepanov, *J. Phys. Chem.*, (1994), 98, 6665.

[23] R.S. Ruoff, R. Malhotra, D.L. Huestis, *Nature (London)*, (1993), 361, 140.

[24] X. Zhou, et. al., *Fullerene Sci. Technol.*, (1997), 5(1), 285.

[25] J. Catalan, J.L. Saiz, J.L. Laynez, N. Jagerovic, J. Elguero, *Angew. Chem. Int. Ed.*, (1995), 34, 1, 105.

[26] R.V. Honeychuck, T.V. Cruger, J.J. Milliken, *J. Am. Chem. Soc.*, (1993), 115, 3034.

[27] K.A. Affholter, S.J. Henderson, G.D. Wignall, G.J. Bunick, R.E. Haufler, R.N. Compton, *J. Chem. Phys.*, (1993), 99, 9224-9229.

[28] H.E. Smorenburg, R.M. Crevecoeur, I.M. de Schepper, L.A. de Graaf, *Phys. Rev. E.*, (1995), 52, 2742.

[29] A.L. Smith, E. Walter, M.V. Korobov, O.L. Gurvich, *J. Phys. Chem.B.*, (1996), 100, 6775.

[30] M.V. Korobov, A.L. Mirakyan, N.V. Avramenko, R.S. Ruoff, *Doklady Akademii Nauk*, (1996), 349, 3, 346, (*In Russian*).

[31] M.V. Korobov, A.L. Mirakyan, N.V. Avramenko, G. Olofsson, A.L. Smith, R.S. Ruoff, *J. Phys. Chem. B*, (1999), 103, 1339.

[32] X. Zhou, J. Liu, Z. Jin, Y. Wu, Z. Gu, J. Yun, B. Wang, Z. Li, Y. Zhang., *Recent Advances in the Chemistry and Physics of Fullerenes and Related Materials* , K.M. Kadish, R.S. Ruoff eds., (1995), 95 10, 1544.

[33] Y.B. Melnichenko, G.D. Wignall, R.N. Compton, G. Bakale, *J. Chem. Phys.*, (1999), 111, 4724.

[34] S.J. Henderson, *Langmuir*, (1997), 13, 6139.

[35] F. Migliardo, V. Magazu, M. Migliardo, *Journal of Molecular Liquids* , (2004), 110, 3-6.

[36] T.V. Tropin, M.V. Avdeev, V.L. Aksenov, *In BNC Progress Report 2002-2003*, (*in press*).

[37] B. Lee, F.M. Richards, *J. Mol. Biol.*, (1971), 55, 379.

[38] T. J. Richmond, *J. Mol. Biol.*, (1984), 178, 63.

[39] A.Yu. Teterev, M.V. Avdeev, Kh.T. Kholmurodov, V.L. Aksenov, *In these Proceedings*.

[40] V.L. Aksenov, T.V. Tropin, M.V. Avdeev, V.B. Priezzhev, J.W.P. Schmelzer, *Kinetics of cluster growth in fullerene molecular solutions* , ECHAYA, (*in press*).

[41] J.W.P. Schmelzer, G. Röpke, J. Schmelzer jn., V.V. Slezov, *In: Nucleation Theory and Applications*, J.W.P. Schmelzer, G. Röpke, V.B. Priezzhev eds., JINR: Dubna, 1999.

[42] A. Beeby, J. Eastoe, R.K. Heenan, *J. Chem. Soc. Chem. Commun.*, (1994), 173.

[43] U.-S. Jeng, T.-L. Lin, C.-T. Tsao, C.-H. Lee, et al., *J. Phys. Chem.*, (1999), 103, 1059.

[44] N.O. Mcheldov-Petrossyan, V.K. Klochkov, G.V. Andrievsky, *J. Chem. Soc., Faraday Trans.*, (1997), 93, 4343-4346.

[45] G.V. Andrievsky, V.K. Klochkov, A. Bordyuh, G.I. Dovbeshko, *Chem.Phys.Lett.*, (2002), 364, 8-17.

[46] M.V. Avdeev, A.A. Khokhryakov, T.V. Tropin, G.V. Andrievsky, et al., *Langmuir*, (2004), 20, 4363-4368.

[47] M.V. Korobov, E.B. Stukalin, N.I. Ivanova, N.V. Avramenko, G.V. Andrievsky, *In: The exciting world of Nanocages and Nanotubes*, P. Kamat, D. Guldi, K. Kadish eds., *Fullerenes V.12, Proc. Series of 201st Meeting of The Electrochem. Soc.*, 2002-12, 799-804, Pennington: NJ, 2002.

[48] Yu.I. Prilutski, S.S. Durov, V.N. Yashchuk, T.Yu. Ogul'chansky, E.V. Pogorelov, Yu.A. Astashkin, E.V. Buzaneva, Yu.D. Kirghisov, G.V. Andrievsky, P. Scharff, *Eur. Phys. J.*, (1999), 9, 341.

[49] A. Mrzel, A. Mertlj, A. Omerzu, M. Eopie, D. Mihailovic, *J. Phys. Chem.*, (1999), 103, 11256.

[50] V.L. Aksenov, M.V. Avdeev, D. Mihailovic, A. Mrzel, V.D. Vasiliev, A.A. Timchenko, I.N. Serdyuk, Study of fullerene aggregates in pyridine/water solutions, *In: Electronic properties of novel materials – molecular nanostructures*, H. Kuzmany, J. Fink, M. Mehring, S. Roth eds., 541, 66, AIP Conference Proceedings, 2001.

[51] V.L. Aksenov, M.V. Avdeev, A.A. Timchenko, I.N. Serdyuk, R.P. May, Aggregation of fullerenes in pyridine/water solutions, *In: Frontiers of Multifunctional Nanosystems*, E. Buzaneva, P. Scharff eds., 281, Kluewer Academic Publishers: Netherlands, 2002.

[52] M.V. Korobov, N.I. Ivanova, N.V. Kojemyiakina, M.V. Avdeev, V.L. Aksenov, *BNC Progress Report*, (2002-2003), (*in press*).

[53] I. Bako, G. Palinkas, J.C. Dore, H. Fischer, P. Jovari, *Chem. Phys. Lett.*, (2004), 388, 468.

[54] A.L. Bulavin, V. Kopylchuk, V. Garamus, M. Avdeev, L. Almasy, A. Hohryakov, *Appl. Phys. A*, (2002), 74, 546.

In: Molecular Simulation Studies in Material ... ISBN 1-59454-607-X
Editor: Kholmirzo Kholmurodov, pp. 129-134

Chapter 10

ORGANIZATION OF SOLVENT AT INTERFACE WITH FULLERENE IN SOLUTIONS C60/CARBON DISULFIDE BY MOLECULAR DYNAMICS SIMULATIONS

A. Yu. Teterev, V. L. Aksenov
Frank Laboratory of Neutron Physics, Joint Institute for Nuclear Research
141980, Dubna, Moscow Region, Russia, Physical Department
Moscow State University, Russia
M. V. Avdeev
Frank Laboratory of Neutron Physics, Joint Institute for Nuclear Research
141980, Dubna, Moscow Region, Russia
Kh. T. Kholmurodov
Department of Radiation and Radiobiological Research,
Joint Institute for Nuclear Research, 141980, Dubna, Moscow Region, Russia

Abstract

The features of the fullerene-solvent interaction in the system fullerene C60/carbon disulfide are analyzed by means of the molecular dynamic simulations. A specific organization of the solvent molecules close to the fullerene surface is revealed. It can affect significantly the visible size of fullerene in this solvent determined by different techniques. The origin of such organization is discussed.

1. Introduction

In recent years fullerene solutions are actively studied. They exhibit some unusual properties such as non-monotonous behavior of temperature dependence of the solubility, solvatochromism and others [1], [2]. The connection of these effects with the possible cluster state of fullerenes are discussed in these Proceedings [3].

The given work concerns the solution of fullerene C60 in carbon disulfide (CS_2), which possesses a comparatively high solubility, ~7.9 mg/ml. The system belongs [4] to a class of solutions where a peak in the temperature dependence of solubility is observed (T_{max} ~

280 K). It has been studied [5]-[11] by different methods to reveal structural features of the fullerene in solution, as well as details of the fullerene-solvent interaction. The reported conclusions on some structural questions are contradictory. In particular, this concerns the size of C60 in CS_2. Thus, according to a number of experimental works on small-angle neutron scattering (SANS) the determined size is about 10 % more than that calculated from the known [12], [13] atomic coordinates of the fullerene molecule. Several assumptions to explain this difference can be made [3]. Here, using molecular dynamics simulations (MDS) we consider the possibility of a specific arrangement of the solvent molecules at the interface with the fullerene. The pure CS_2 is well studied by means of MDS [14]. The idea of the work is based on the fact that the characteristic size of the CS_2 molecule (~0.3 nm) is comparable with that of fullerene C60 (~1 nm), so that any interface organization of the CS_2 molecules different from that in bulk must result in a significant difference between the interface and bulk molecular density of the solvent, and, hence, affects the visible size of the fullerene. Below, we show that really the expected difference takes place in the considered solution. Its origin is discussed.

2. Object of Investigation and Method

MDS calculations were performed with the program DL POLY [15]. First, they were made for the pure solvent basing on the work [14]. Then, after comparison of the results, they were repeated for the fullerene solution.

Initially, the basic cube in the calculations of 37.82 Å edge contained 512 molecules of CS_2. Toroidal boundary conditions were employed. The starting configuration was a simple cubic lattice of carbon atoms with randomized molecular orientations. Initial translations and rotational velocities were set equal to zero. Throughout the computation, CS_2 molecules were treated as rigid with bond length of 1.55 Å between carbon and sulfide atoms. Only the Lennard – Jones interaction were taken into account:

$$u_{AB}(r_{AB}) = 4\varepsilon_{AB}\left[\left(\frac{\sigma_{AB}}{r_{AB}}\right)^{12} - \left(\frac{\sigma_{AB}}{r_{AB}}\right)^{6}\right], \tag{1}$$

where $u_{AB}(r_{AB})$ is the interaction energy between atoms A and B localized at the distance of r_{AB} from each other; ε_{AB} and σ_{AB} are the interaction constants. Because of the symmetry of the CS_2molecule only three types of interaction are to be specified for ε_{AB} and σ_{AB}, which correspond to bonds sulphur–sulphur, carbon–sulphur and carbon–carbon. The actual values used in the calculations were taken from the previous work [14]:

$\varepsilon_{SS} = 0.77$ kcal/mol $\sigma_{SS} = 3.3$ Å
$\varepsilon_{CS} = 0.06$ kcal/mol $\sigma_{CS} = 4.0$ Å
$\varepsilon_{CC} = 0.06$ kcal/mol $\sigma_{CC} = 3.85$ Å.

The forces and torques at a given molecule were summed in pairs up to cutoff radius $R_c = 10$ Å. The number density of liquid was taken to be $\rho = 1.2$ g/cm^3, which corresponds to the mean distance between carbons atoms of 4.73 Å. Time-step was set as $\Delta t = 10^{-4}$ ps. The whole computer run was taken 2870000 and 640000 turns for the first and second calculations, respectively. Temperature was stabilized with the Hoover thermostat at 293K.

For the fullerene solution the basic cube length were expanded up to 104 Å with the corresponding increase in a number of CS_2 molecules up to 10680. The central molecule was changed by the molecule of C60 fullerene, which corresponded to the fullerene concentration of about 1.2 mg/cm^3. The bonds in the fullerene molecule were considered as rigid with the lengths of 1.39 and 1.44 Å [16]. Other parameters of the calculations were the same like in the previous case [14].

3. Results and Discussion

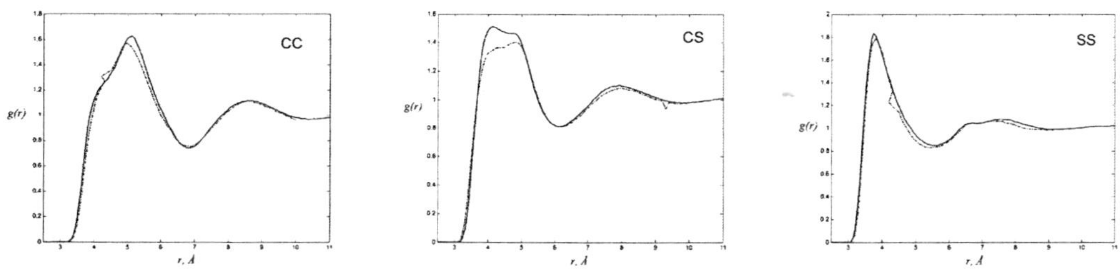

Figure 1. Graphs of the $g(r)$ function for the different types of correlation in pure CS_2 calculated in [14] (dashed lines) and in the current work (solid lines)

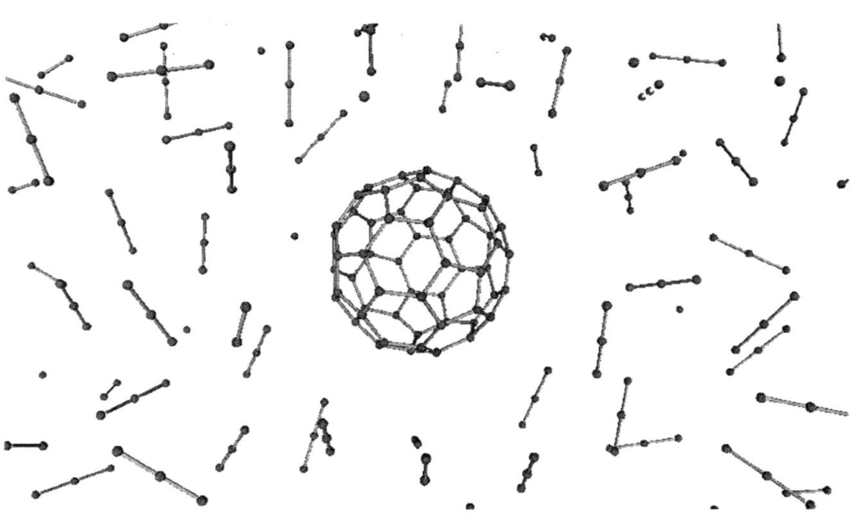

Figure 2. Instantaneous slice of the system C60/CS_2 close to the interface between fullerene and solvent

The comparison of the resulting $g(r)$ functions for the pure CS_2 with those obtained previously [14] is given in Fig.1. It shows good agreement between both calculations. A slice of the molecular configuration around fullerene calculated for the solution C60/CS_2 is presented in Fig.2. To analyze the organization of the solvent molecules at the interface with fullerene the concentration function, $n(r)$, for two kinds of atoms was calculated with the origin in the center of mass of the fullerene molecule. The results are shown in Fig.3.

One can see that a specific modulation of the atomic concentration takes place close to the fullerene surface. It is similar by nature to a short order seen in Fig.1. In experiments on the scattering from liquids the influence of such modulation takes place at large values of the momentum transfer (wide-angle diffraction) because of the small distances. But in the case of fullerene solution, the corresponding density fluctuations are connected with the size of fullerene molecule, and, hence, should contribute to the coherent scattering from fullerene in the SANS signal. The calculated scattering length density as a function of the distance from the center of fullerene molecules is plotted in Fig.4. The bulk value of the density at large r-values is about 1.20×10^{10} cm^{-2}, which is in good agreement with the density calculated from the physical solvent density, $\rho = 1.22\times10^{10}$ cm^{-2}.

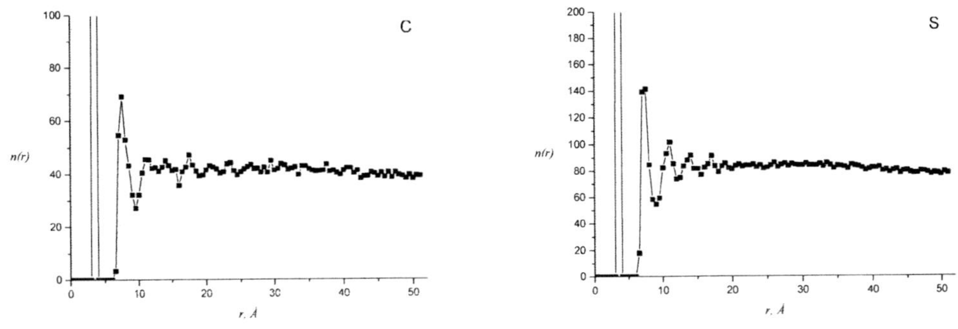

Figure 3. Graphs of the $n(r)$ function for the different types of solvent atoms at the interface between CS_2 and C60

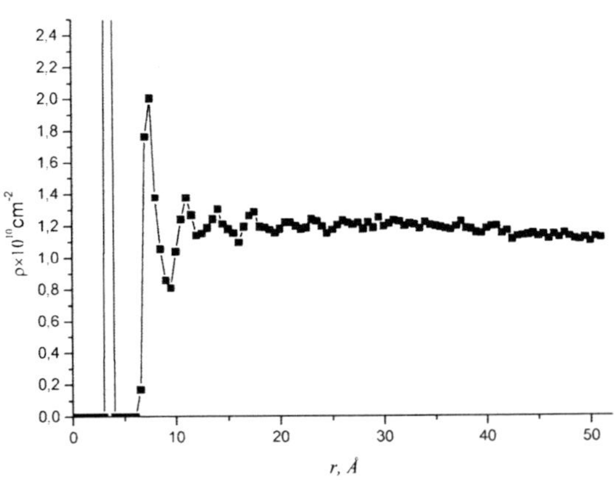

Figure 4. Profile of the scattering length density close to fullerene surface in solution of C60 in CS_2

4. Conclusion

So, one can conclude that according to MDS data a special organization of the solvent molecules on the interface in the solution C60/CS_2 takes place. It is of the short order type similar to that in usual liquids. The reason why it should be reflected in the experiments on small angle scattering is connected with a comparatively large size of the fullerene molecule. The obtained modulation of the scattering length density at the interface of CS_2 with fullerene should be taken into account when calculating the scattering curve from fullerene and its visible size in scattering experiments.

Acknowledgments

The work has been performed with the support of the Russian Ministry Education and Science, state contract №40.012.1.1.1148.

References

[1] V.N. Bezmelnitsyn, A.B. Eletskii, M.B. Okun, *Usp. Fiz. Nauk*, (1998), 168, 1195 (*in Russian*).

[2] M.V. Korobov, A.L. Smith, *In: Fullerenes: chemistry, physics and technology*, 53, K.M. Kadish, R.S. Ruoff eds., A John Wiley&Sons, Inc., 2000.

[3] M.V. Avdeev, T.V. Tropin, V.L. Aksenov, *In: these Proceedings*.

[4] R.S. Ruoff, R. Malhotra, D.L. Huestis, *Nature (London)*, (1993), 361, 140.

[5] K.A. Affholter, S.J. Henderson, G.D. Wignall, G.J. Bunick, R.E. Haufler, R.N. Compton, *J. Chem. Phys.*, (1993), 99, 9224.

[6] Y.B. Melnichenko, G.D. Wignall, R.N. Compton, G. Bakale, *J. Chem. Phys.*, (1999), 111, 4724.

[7] H.E. Smorenburg, R.M. Crevecoeur, I.M. de Schepper, L.A. de Graaf, *Phys. Rev. E.*, (1995), 52, 2742.

[8] S.J. Henderson, *Langmuir*, (1997), 13, 6139.

[9] F. Migliardo, V. Magazu, M. Migliardo, *J. Mol. Liq.*, (2004), 110,3.

[10] T. Tomiyami, S. Uchiyama, H. Shinohara, *Chem. Phys. Lett.*, (1997), 264, 143.

[11] A.D. Bokare, A. Patnaik, *J. Chem. Phys.*, (2003), 119, 4529.

[12] R.M. Fleming, et al., *Mater. Res. Symp. Proc.*, (1991),206, 691.

[13] D.R. Mckenzie, C.A. Davis, D.J.H. Cockayne, D.A. Muller, M. Vasssalo, *Nature*, (1992), 355, 622.

[14] O. Steinhauser, M. Neumann, *Mol. Phys.*, (1979), 37, 1921.

[15] http://www.cse.clrc.ac.uk/msi/software/DL_POLY

[16] A.B. Eletskii, B.V. Smirnov, *Usp. Fiz. Nauk*, (1993), 163, 33 (*in Russian*).

In: Molecular Simulation Studies in Material ... ISBN 1-59454-607-X
Editor: Kholmirzo Kholmurodov, pp. 135-146

Chapter 11

MD Simulations on a Flexible Chain Polyethilene: The Effect of the Electrostatic Forces in the Density-Temperature Behavior

D. A. Kretov and Kh. T. Kholmurodov*
Center for Molecular Modeling, Department of Radiation and Radiobiological Research, Joint Institute for Nuclear Research
141980, Dubna, Moscow Region, Russia

Abstract

Molecular Dynamics (MD) simulations have been performed on the parallel-vector and special-purposes MD machines to study the dynamical behavior of a flexible polyethylene (PE) chain subjected to various heating conditions. Both the effect of the short-range and electrostatics interactions on the dynamics of the PE system was estimated and compared. Main objective of the present study was to investigate the influence of the heating condition on a final state of the polymeric system and sensitivity of its thermodynamic quantities (density, energy, etc.) on the heating modes. For this purpose the three different simulations were performed regarding on the following regimes: fast, moderate, and slow heating processes. The peculiarities of the main thermodynamic quantities of the system were estimated in detail and compared between all three processes. Molecular mechanical and MD calculations on the PE system were performed using the optimized version of the DL_POLY code, adapted for the vector- and special-purpose MDGRAPE-2 machines.

1. Introduction

One of the most convenient techniques to study polymer phase transitions and other technologically interesting phenomena (structural conformations, chain-folding processes, time

*E-mail address: kretov_da@mail.ru, mirzo@jinr.ru

and density-temperature behaviors, etc.) is molecular dynamics (MD) method. Simulations on the detailed structural changes of polymers, based on their microscopic transition behavior, provide the information valuable for the industry of polymer materials, for the material modification and technological innovations [1]-[6]. The MD simulation allows one to estimate the density-temperature changes of polymer systems, to study their time dependent thermodynamic properties in solvents and to explore polymer-surface interactions [1]-[5],[7]. In this work we have studied through computer MD simulations the density-temperature distribution and dynamical properties of a chain PE, subjected to various heating conditions. It is worth noting that for polymeric systems both temperature and density play crucial roles in their dynamics and formation of nontrivial phases. For example, it has long been appreciated that both temperature and density essentially determine the low-temperature behavior of polymeric chains (the extremely super-Arrhenius phase), the viscosity of the fragile liquids, etc. [8]-[10]. In several recent studies for a model polymeric material (see, for example, [8]-[13]) the computer simulations were performed on the basis of the MD to study the chain dynamics in bulk PE over a wide range of temperature (including the volumetric glass transition), the isobaric and isochoric cooling and other phenomena. It was found that simulation results, in general, are in excellent agreement with the existing experimental trends [10]-[12]. As a main objective of the present study we have investigate the influence of the heating condition on a final state of the polymeric system and sensitivity of the thermodynamic quantities (density, energy, etc.) on the heating processes. The MD simulations we have performed on the basis of the optimized general-purposes DL_POLY code [14],[15] and generic DREIDING force field for molecular simulations [16].

2. Details of Simulations

The MD simulations of a single flexible polyethylene (PE) chain of about 500 CH2 units in length (each CH2 group is treated as a united atom) were performed using DREIDING force field potential. A snapshot of the initial configuration of the PE chain is shown in Fig.1.

The initial PE configuration was constructed with the help of the built-in interface provided by the DL_POLY package, supporting a number of the utilities for the analyzing of the results and visualization purposes.

The configuration energy of the PE model is represented as a sum of the intramolecular valence (bonded) E_{val} and nonbonded E_{nb} interaction energies:

$$E = E_{val} + E_{nb}. \tag{1}$$

The intramolecular energy E_{val} includes bond stretching, bond angle bending, and torsion and inversion interactions:

$$E_{val} = E_R + E_A + E_T + E_I, \tag{2}$$

nonbonded interactions consist of Van der Waals (VdW), electrostatic and hydrogen bonds:

$$E_{nb} = E_{vdw} + E_Q + E_{hb}. \tag{3}$$

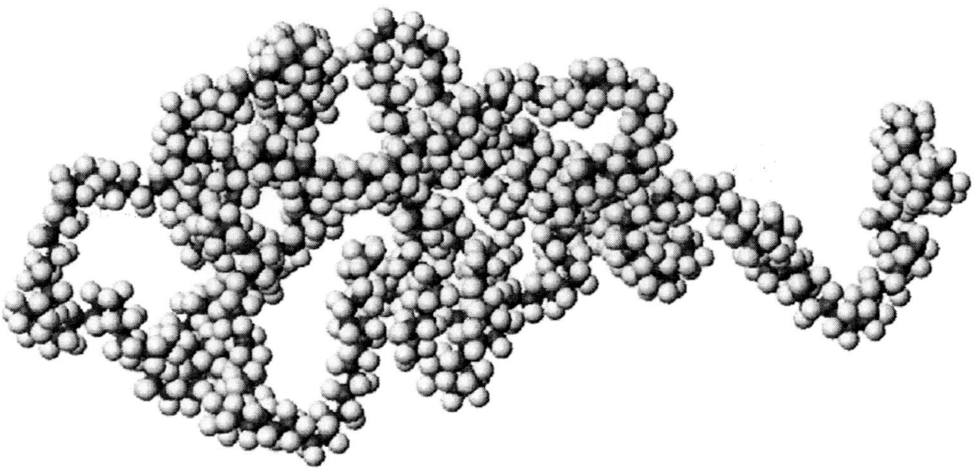

Figure 1. Initial configuration of a polyethylene chain (black circles: carbon atoms, white circles: hydrogen atoms

Table 1. The LJ-parameters used in the MD simulations.

Atomic pair	$\varepsilon, kcal/mol$	$\sigma, \AA$
C C	0,0951	3,4730
C H	0,0380	3,1597
H H	0,0152	2,8464

The VdW interactions were calculated by the pairwise 6-12 Lennard-Jones (LJ) potential. The LJ-parameters of the PE model are shown in Table 1.

We employed the NPT ensemble; a constant temperature of 300 K was attained using Nose-Hoover method. The equation of motion was solved by the leapfrog Verlet algorithm with a time step of 0.001 ps. The relaxation constant for the heat bath variable was 0.1 ps. In Table 2 the control parameters used in the MD simulations are shown.

3. Results and Discussion

We have studied the influence of the heating processes on a final state of the PE system with periodic boundary conditions. Starting from the same initial state at $T_0 = 10K$ with the system equilibration we have reached a final state of $T_F = 300K$ through the following three different ways: regime 1 - fast, regime 2- moderate and regime 3 - slow heating modes. In Table 3 the temperature steps for those three simulation regimes are displayed.

For each heating mode described above all atomic interactions were estimated through the following stages:

1. The valence and VdW forces calculated. The contribution of the electrostatic interactions at this step was neglected.

Table 2. The setting control parameters of the MD simulations.

Temperature, K	300
Pressure, kbar	0
Ensemble	NPT
$R_{cut}, \dot{A}$	10,5
Ewald precision	10^{-4}

Table 3. The temperature increments for the three heating regimes.

Heating regime	Temperature increment $\Delta T, K$
Fast	150
Moderate	50
Slow	10

2. The valence, VdW and the electrostatics forces calculated.

Thus, the effect of the electrostatic interactions on the dynamics and density-temperature behavior of the PE system was compared to the valence and VdW interactions.

3.1. The Effect of the Valence and VdW Interactions ($the electrostatics = 0$)

In Fig.2 the values of the temperature (T) vs time (t) are shown for three regimes: left picture displays fast regime, middle picture - moderate, and right picture - slow heating. The starting time $T = 4ps$ coincides with the end of the equilibration process. In each simulation run several thousands MD steps with the velocity scaling were performed to reach the equilibrium states, viz., the point $t = 4ps$.

It is easily seen from the Fig.2 that the evolution of the temperature has a clearly different behavior between the heating regimes. During the fast and moderate heating modes we observe large oscillations of the temperature ($T(t)$). For the same system, the slow heating yields satisfactory small oscillations of temperature. In other words, the slow heating mode "smoothly" forms the equilibrium state, which manifests itself as an important stage of the final dynamics. A final state achieved at $140ps$ has the same value of T close for all heating regimes. For a larger MD time scales ($t > 140ps$) a more significant changes of the temperature values has not been observed, so far the PE system has been analyzed only at this time interval from 0 to 120 picoseconds.

Next we have estimated the evolutions of the atomic density of the system for the three heating processes. In Fig.3 the behavior of $\rho_N(t)$ is shown for the same regimes as displayed in Fig.2 (left picture: fast; middle: moderate; and right: slow heating).

In the fast and moderate heating the visibly high oscillations of the density values were observed. In comparison to the pictures above for time dependences of $T(t)$, the high temperature oscillations result in a significant change of the density (see left and middle

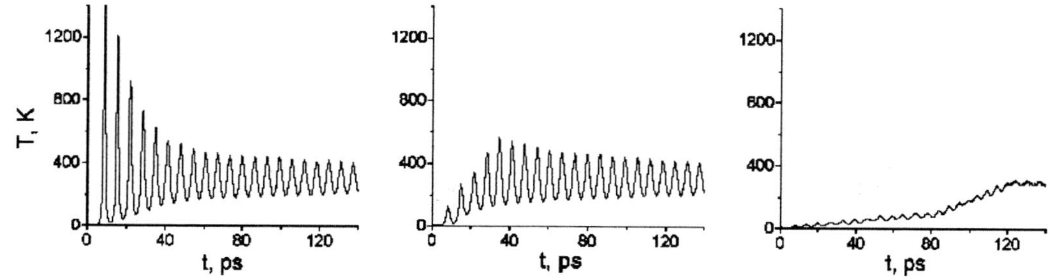

Figure 2. The evolutions of the temperature of the system in the fast (left), moderate (middle) and slow (right) heating regimes.

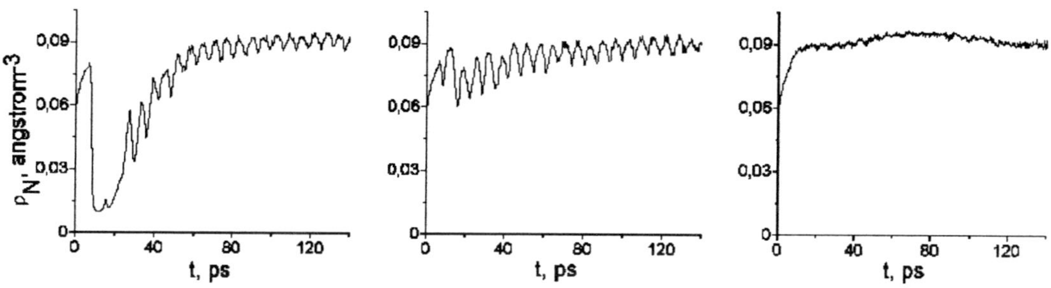

Figure 3. The evolutions of the atomic density in the fast (left), moderate (middle) and slow (right) heating regimes.

pictures). At the same time, the slow heating process leads to an equilibrium density monotonically. That is the behavior of the system under consideration, as it seen from Fig.3, is essentially sensitive to the dynamics of the heating. The values of $rho_N(t)$ at the final point $t = 140ps$ are the same for all heating regimes. The final equilibrium values for the three regimes are presented in Table 4.

The various energy contributions, which result in the density behavior as shown above, have been compared. We have separately estimated the energies of the non-bonded (VdW) and intramolecular (bonded) interactions for all heating regimes. In Fig.4 the calculated values of the configuration energy due to the short-range potential E_{wdv}, as a function of time, are presented.

It is seen from Fig.4 that the changes of at the beginning of the MD simulations for the fast and moderate heating processes (left and middle pictures) are approximately one order higher than those for the slow heating mode (shown in the right picture). Starting from the same value of about $E_{vdw} = 0,125 \times 10^3 kcal/mol$ the energy grows up to $1,125 \times$

Table 4. The calculated densities of the PE system for the three heating regimes.

Regime	Fast	Moderate	Slow
$\rho, g/cm^3$	0,165532150	0,160132	0,165256

$10^3 kcal/mol$ in regime 1 (left picture) and up to $0,75 \times 10^3 kcal/mol$ in regime 2 (middle picture), respectively. At the same time, as it seen from Fig.4, the VdW energy for regime 3 (right picture) keeps about the average value of $E_{vdw} = 0,125 \times 10^3 kcal/mol$.

In Figs.5-7 we present the values of the intramolecular energies due to the bond E_B (Fig.5), angular E_A (Fig.6), and dihedral E_T (Fig.7) interactions as a function of time.

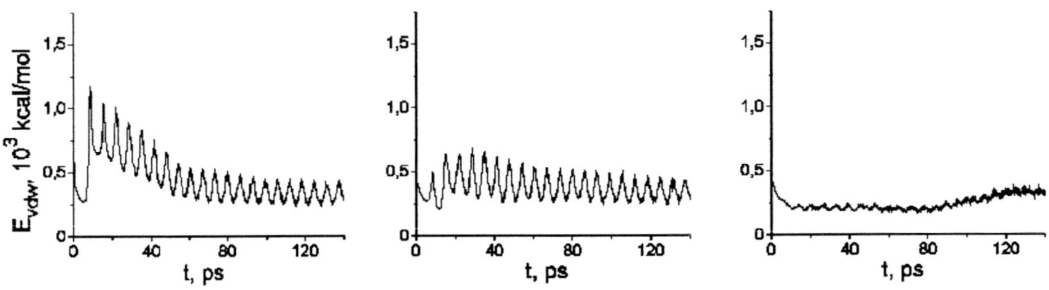

Figure 4. Configuration energy due to the short-range (VdW) potential in the fast (left), moderate (middle) and slow (right) heating regimes.

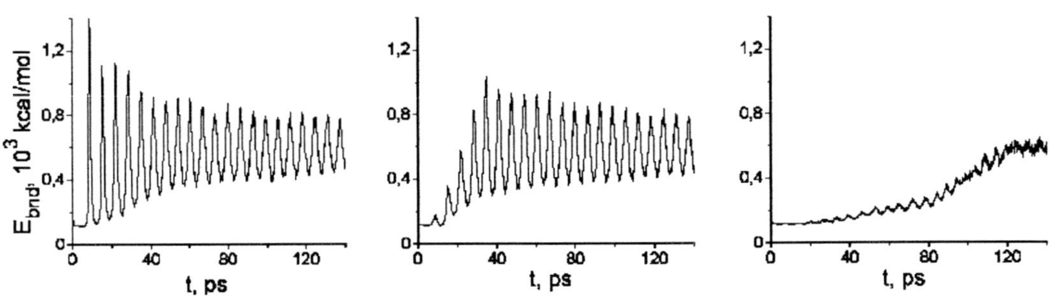

Figure 5. Configuration energy due to intramolecular (chemical bond) potential in the fast (left), moderate (middle) and slow (right) heating regimes.

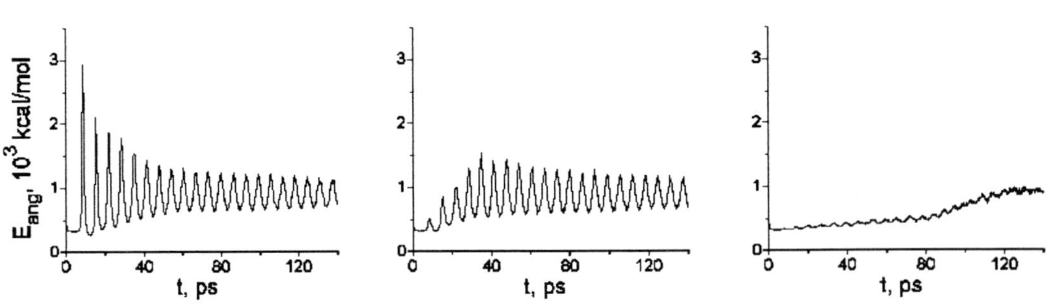

Figure 6. Configuration energy due to intramolecular (valence angle) potential in the fast (left), moderate (middle) and slow (right) heating regimes.

As it seen from the Figs.5-7 the behavior of the energy for the bond, angular and dihedral interactions could easily be distinguished, comparing those for the three heating regimes. For the fast and moderate heating modes the initial changes of the intramolecular energies are significantly larger than those for the slow heating process. The oscillations of the intramolecular energies for the slow heating behave monotonically. This is in accordance with the dynamics of the setting temperature as shown in Fig.2. The chemical bond potential gives a large contribution; next is the valence angle potential; and then goes the dihedral interaction potential. It is worth noting that the changes of the dihedral energy in slow heating mode are very small. This is an interesting dynamical behavior to be taken into account in similar studies performed on the PE system. The snapshots of the resultant configurations of the PE system for the three heating modes are presented in Fig.8, respectively.

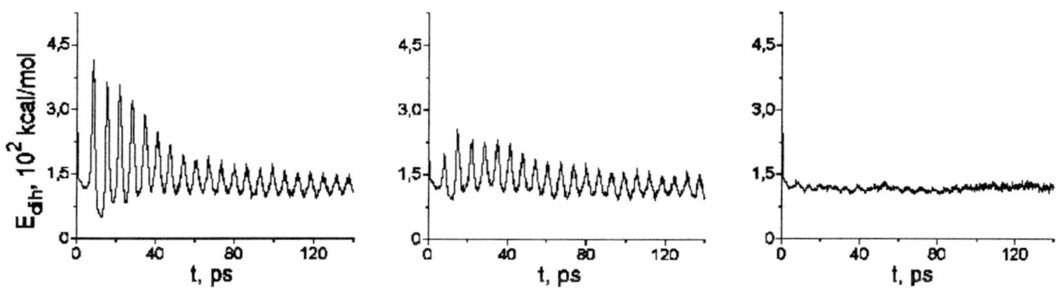

Figure 7. Configuration energy due to intramolecular (dihedral) potential in the fast (left), moderate (middle) and slow (right) heating regimes.

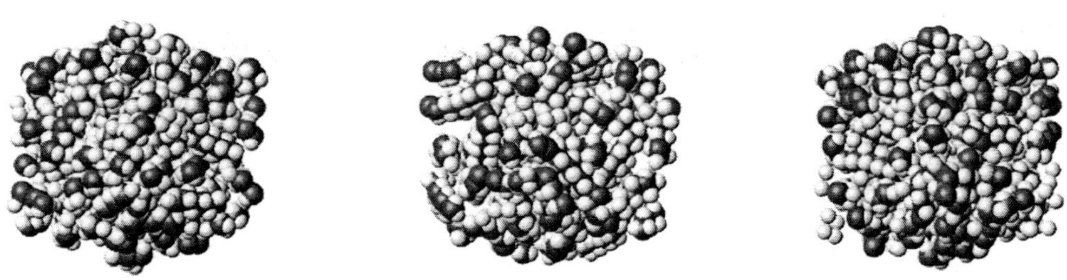

Figure 8. Snapshots of the resultant configurations of the PE system for the fast heating (left), moderate (middle) and slow (right) regimes.

Finally, in Fig.9 the results of the total energies of the PE system for the regimes 1 (left), 2 (middle) and 3 (right) are displayed.

From Fig.9 we see that all displayed values of the total energy asymptotically go to a constant. Nevertheless, it is important that those values calculated for the PE system differ from each other. The equilibrium energy E_{tot} is $0,6 \times 10^3 kcal/mol$ for the regime 1; $1,3 \times 10^3 kcal/mol$ for regime 2; and $0,9 \times 10^3 kcal/mol$ for regime 3, respectively. The most remarkable point, however, is that in the fast and moderate heating the total energy behave not in the same manner, as this clearly was observed from the results of Figs.2-7. In comparison to the above dependencies (see, for example, Figs.4-7 for the nonbonded and the bonded potentials, respectively) the total energy starting at $t = 0ps$ from the close values (Fig.9) decreases in the fast, but increases in the moderate heating mode (before reaching an asymptotics). This can rather be explained so that the total energy E_{tot}, as a conserved quantity of the system, contains all contributions including those coming from

Table 5. The hydrocarbon mass and charge numbers.

Atomoc index	Mass	Charge
C	12,011000	-0,053302
H	1,008000	0,026651

the thermostat. Thus, the PE system interacting with the thermostat forces reaches the different values of E_{tot} (see Fig.9 for the regimes 1 and 2). This is true even when the intra- or intermolecular energy parts possess the same law, comparing the fast and moderate regimes of Figs.4-7. In other words,the thermostat contribution somehow "corrects" the final state, yielding the correct constant values of E_{tot}.

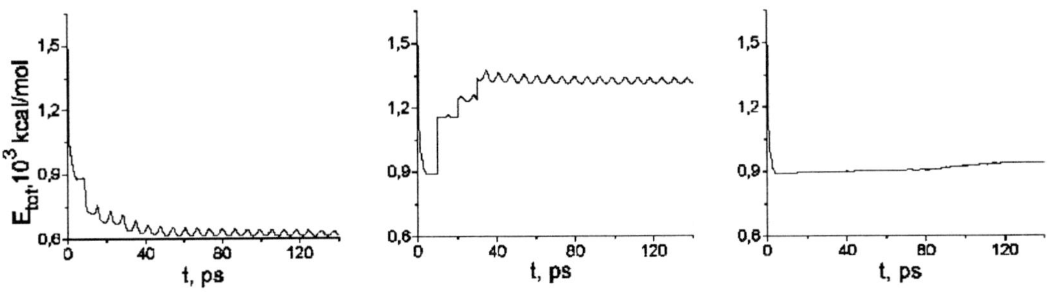

Figure 9. The evolution of the total energy of the system for the fast (left), moderate (middle) and slow (right) heating regimes.

3.2. The Effect of the Valence and VdW Interactions (*the electrostatics* $\neq 0$)

In this section we present the results of all potential type calculations, including contribution of the electrostatic forces, and then compare the density-temperature dynamical chnges of the PE system with those described in Figs.2-9. Within the DREIDING force field the CH_2 group of PE chain is represented as united atom, satisfying the charge neutrality. Table 5 displays the mass and charge numbers of the PE model, used in the calculation of the Coulomb interactions.

In Fig.10 the results of the Coulomb energy calculations for the three heating regimes are presented (left picture: fast; middle: moderate; right: slow process).

The electrostatic energy of the PE chain, as it is seen from Fig.10, possesses very small values. In comparison with the valence or VdW potentials (see, Figs.4-7) contribution of the electrostatics is three orders smaller. Nevertheless, this small contribution changes the behavior of basic quantities (temperature, density, energy) essentially. Below we discuss the peculiarities of the temperature, density, and energy evolutions in the same manner as in the previous section 3.1..

In Figs.11 and 12 the temperature and density for three heating modes in the presence of the electrostatic interactions are presented vs time.

From Fig.11 we observe small changes of the temperature dynamical behaviors, if to

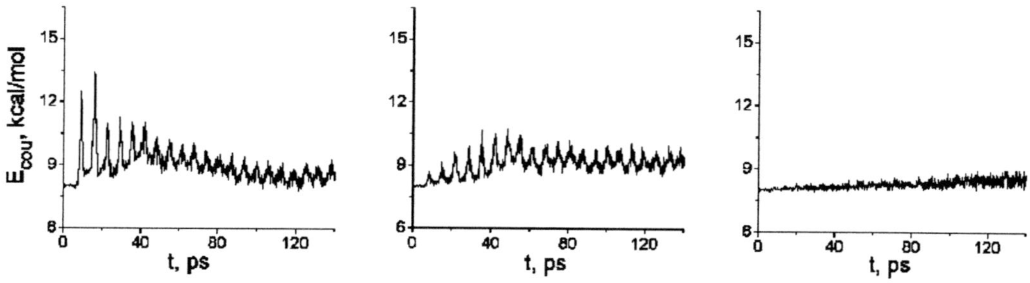

Figure 10. The electrostatic (columb) energy of the PE system for the fast (left), moderate (central) and slow (right) heating regimes.

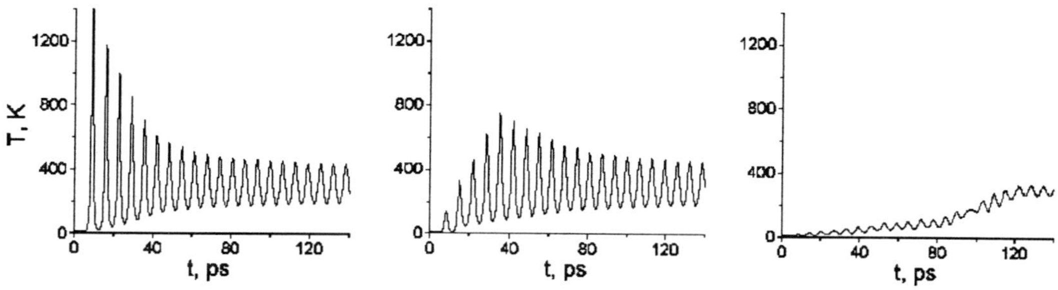

Figure 11. The evolutions of the temperature of the PE system with electrostatics in the fast (left), middle (central)and slow (right) heating regimes.

compare these results with those in Fig.2. The peculiarities of the time evolutions of the temperature are easily distinguishable for the fast and moderate heating; for the slow heating mode the results are closely the same.

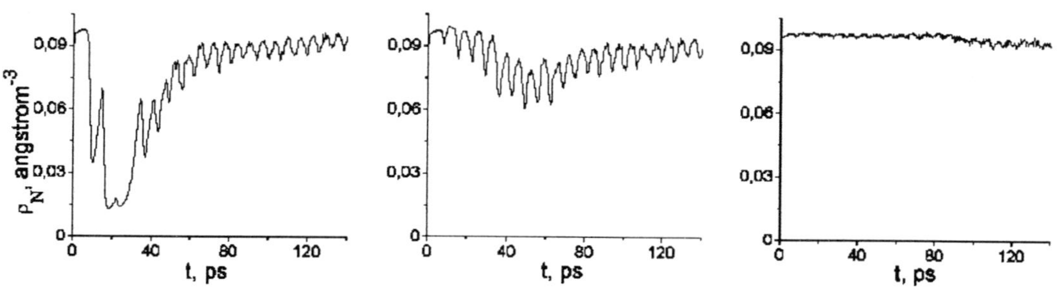

Figure 12. The evolutions of the atomic density of the PE system with electrostatics in the fast (left), middle (central) and slow (right) heating regimes.

In comparison with the behaviors of the temperature (viz., Fig.11) inclusion the electrostatic forces for the density leads to its nonlinear changes. From Fig.12 it is seen that in the moderate and slow heating (middle and right pictures) the density reaches its maximum value very fast, from the start of the evolutions. Besides of this, the density oscillations in the moderate heating pass the extremum at later stages of the systems dynamics. This result represents itself intriguing and it might be interpreted as a PE intermediate phase or in terms of the nonlinear dynamical transitions in the polymeric chain.

The MD calculation results for the VdW and intramolecular energies are shown in Fig.13 and Fig.14-16, respectively.

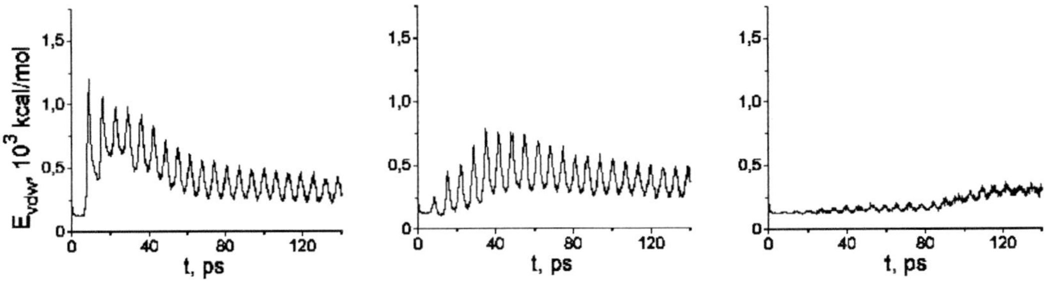

Figure 13. Configuration energy due to the short-range (VdW) potential of the PE system with electrostatics in the fast (left), middle (central) and slow (right) heating regimes.

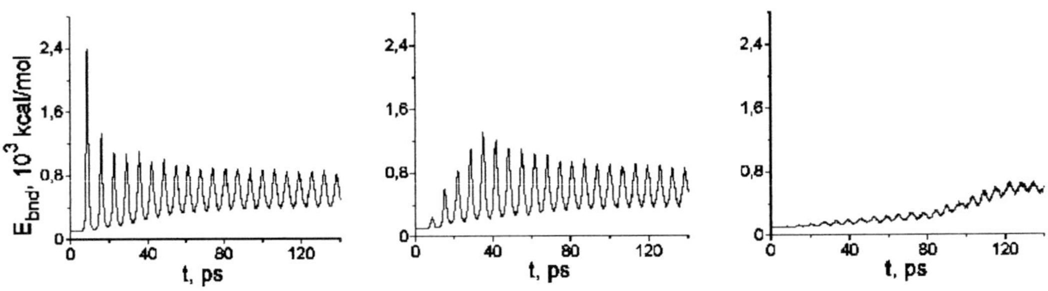

Figure 14. Configuration energy due to intramolecular (chemical bond) potential of the PE system with electrostatics in the fast (left), middle (central) and slow (right) heating regimes.

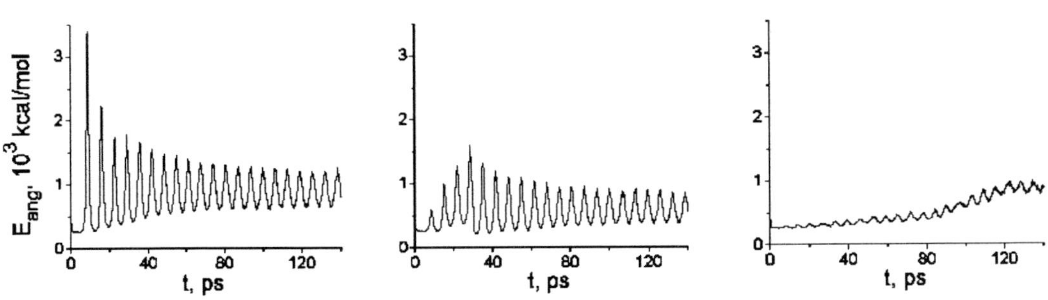

Figure 15. Configuration energy due to intramolecular (valence angle) potential of the PE system with electrostatics in the fast (left), middle (central) and slow (right) heating regimes.

Finally, in Fig.17 the results of the total energies of the PE system with electrostatics for the regimes 1 (left), 2 (middle) and 3 (right) are displayed.

The analysis of the results shown in Figs.13-16 and those in Figs.4-7 indicate that the effects of the electrostatic forces have seen for all heating modes. A largest change more obvious, however, for regime 2 (the middle pictures in Figs.13-16). That is easy to observe that the evolution curves of the moderate heating mode possess a nonlinear region, where VdW or bonded energy goes through a maximum value. In this temporal window (from 30*ps* until 70*ps*) all interaction potentials weakly increased simultaneously with the decreasing of the density.

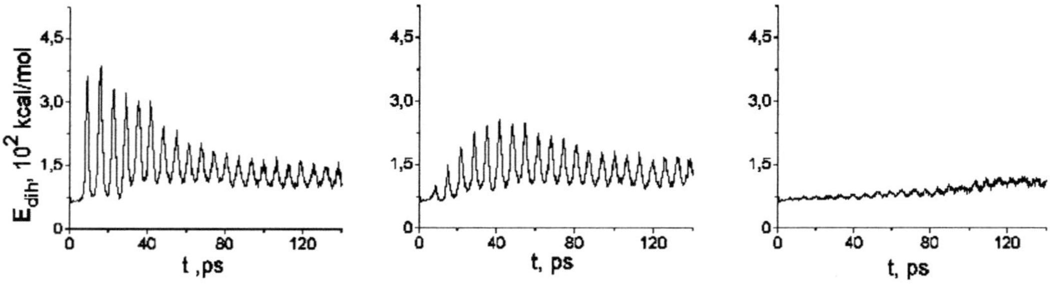

Figure 16. Configuration energy due to intramolecular (dihedral) potential of the PE system with electrostatics in the fast (left), middle (central) and slow (right) heating regimes.

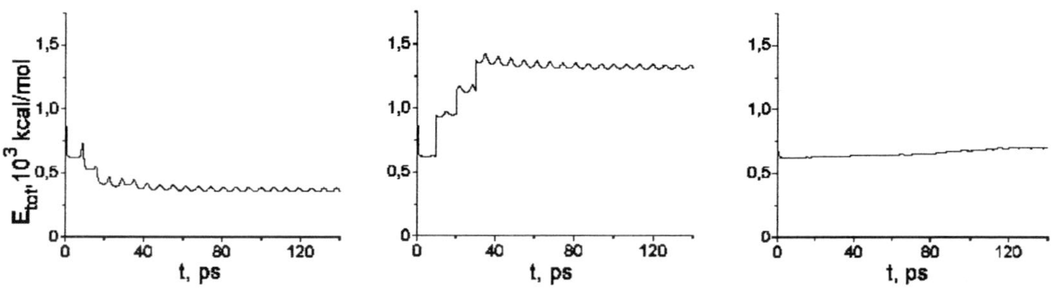

Figure 17. The evolution of the total energy of the PE system with electrostatics for the fast (left), moderate (middle) and slow (right) heating regimes.

Acknowledgments

The special-purposes MDGRAPE-2 computer was provided by Ebisuzaki Computational Astrophysics Laboratory of RIKEN (The Institute of Physical and Chemical Research), Japan. The authors are grateful to Prof. Toshikazu Ebisuzaki for the supports on the MDGRAPE-2 machine and DL_POLY optimization works. We acknowledge the helpful discussion and valuable comments by Dr. Michail Altaisky.

References

[1] N.L. Allinger, *J. Comput. Chem.*, (1987), 8, 581.

[2] S.M. King and T. Cosgrove, *Macromolecules*, (1993), 26, 5414.

[3] J. Budzien, C. Raphael, M.D. Ediger, and J.J. de Pablo, *J. Chem. Phys.*, (2002), 116, 8209.

[4] M. Fukuda and H. Kikuchi, *J. Chem. Phys.*, (2000), 113, 4433.

[5] H.X. Guo, X.Z. Yang, and T. Li, *Phys. Rev. E*, (2000), 61, 4185.

[6] T.K. Xia and U. Landman, *Science*, (1993), 261, 1310.

[7] P.V. Krishna Pant, J. Han, G.D. Smith, and R.H. Boyd, *J. Chem. Phys.*, (1993), 99, 597.

[8] R.H. Boyd, R.H. Gee, J. Han, and Y. Jin, *J. Chem. Phys.*, (1994), 101, 788.

[9] M.L. Ferrer, C. Lawrence, B.G. Demirjian, D. Kivelson, C. Alba-Simionesco, and G. Tarjus, *J. Chem. Phys.*, (1998), 109, 8010.

[10] L. Yang, D.J. Srolovitz, and A.F. Yee, *J. Chem. Phys.*, (1999), 110, 7058.

[11] R.K. Bharadwaj and R.H. Boyd, *J. Chem. Phys.*, (2001), 114, 5061.

[12] A. Koyama, T. Yamamoto, K. Fukao, and Y. Miyamoto, *J. Chem. Phys.*, (2001), 115, 560.

[13] A. Neelakantan, and J.K. Maranas, *J. Chem. Phys.*, (2004), 120, 465.

[14] W. Smith, T.R. Forester, *Molecular Graphics*, (1993), 14, 136; The DL_POLY User Manual, Version 2.11, http://www.dl.ac.uk/TCS/Software/DL_POLY, Daresbury Laboratory, United Kingdom, 1999.

[15] Kh. Kholmurodov, W. Smith, K. Yasuoka, and T. Ebisuzaki, *Comput. Phys. Commun.*, (2000), 125, 167; Kh. Kholmurodov, et al., "Fast Forces Calculations in the DL_POLY Molecular Simulation Package", *In: Proceedings "6th World Multiconference on Systemics, Cybernetics and Informatics (SCI)"*, Orlando, USA, July 14-18, 2002.

[16] S.L. Mayo, B.D. Olafson, and W.A. Goddard III, DREIDING: A Generic Force Field for Molecular Simulations, *J. Phys. Chem.*, (1990), 94, 8897.

In: Molecular Simulation Studies in Material ...
Editor: Kholmirzo Kholmurodov, pp. 147-159 ISBN 1-59454-607-X

Chapter 12

MULTISCALE IMPLEMENTATION OF FAST MULTIPOLE METHOD

Mikhail Altaisky, Victor Krylov and Kholmirzo Kholmurodov*
Center for Molecular Modeling
Department of Radiation and Radiobiological Research
Joint Institute for Nuclear Research
141980, Dubna, Moscow Region, Russia

Abstract

We describe an object-oriented C++ implementation of the three-dimensional Fast Multipole Method (FMM) for evaluation of the Coulombic potential and the force field for large systems of charged or gravitating particles. The FMM is the preferable alternative for the evaluation of the Coulombic forces in the molecular dynamic simulations of large molecules. In contrast to the direct summation $\sum_{j\neq i}\frac{1}{r_{ij}}$, that scales as $O(N^2)$ with the number of particles, the FMM method has $O(N\log N)$ asymptotic. The FMM algorithm is extremely important for simulation of large biomolecules, first of all for proteins and nucleic acids, and for the simulation of large gravitating systems is astrophysics. The implementation have been performed according to the Greengard-Rokhlin algorithm [*Acta Numerica*, 1997]. The code was tested on ordinary PC for a systems of up to 10^6 particles.

1. Introduction

The evaluation of the electrostatic interaction between atoms and molecules is the most time consuming part of many problems of the dynamics of large discrete systems. This happens in simulation of the dynamics of galaxies, driven by the $U = G\frac{m_1 m_2}{R}$ gravitational interaction, in electrostatic interaction of charge particle

$$U = \sum_{i\neq j}\frac{q_i q_j}{|r_i - r_j|}, \tag{1}$$

*E-mail address: [altaisky,kryman,mirzo]@jinr.ru

in $1/R$ interaction of vortices in a fluid flow [1], and in many other problems. For a small system the sum 1 is easy to evaluate by direct summation. For a large systems, say $N \geq 10^4$ particles, this becomes a huge computational problem, for the time of calculation grows as $O(N^2)$ with the number of particles. Since the gradient of potential – the force field – is evaluated at each time step for each particle in the Newtonian dynamics simulation, the calculation

$$\begin{aligned} v_i(t+\Delta t) &= v_i(t) - \Delta t \frac{1}{m} \nabla_i U + \ldots \\ r_i(t+\Delta t) &= r_i(t) - \Delta t v_i(t) + \ldots, \end{aligned}$$

requires millions of operations at each time step. Therefore the direct numerical simulation of a real protein or a DNA structure of moderate size still remains out of the capabilities of modern supercomputers.

To overcome this obstacle a number of fast algorithms and special hardware solutions for the Coulombic interaction have been developed. Among those the Evald method based of the FFT is of most often use [2]. The time consumption of the Ewald method scales approximately as $O(N \log N)$. This is reasonably efficient and is implemented in special-purpose MDGRAPE hardware and software for the molecular dynamic simulations [3, 4]. Starting from the Barnes-Hut multiscale algorithm [5], the $O(N \log N)$ scaling of resource consumption is a typical feature of multiscale algorithms – FFT, wavelet and lifting schemes – that calculate the final result by sequential coarse-graining of the system, with the near neighbors being accounted locally, and the far neighbors being accounted from the next scale.

The theoretical limit of the resource consuming of multiscale algorithms $O(N \log N)$ can be improved a little toward $O(N)$ asymptotic [6, 7]. However, an effective 3D implementation of the multiscale structures is a subtle question, that requires not only the optimization of mathematical methods, but also taking into account the particular type of cashing and interprocessor communications. Therefore, it is important to tune the parameters used for particular simulation – number of particles, density distribution, etc. – that can be done only in software realization.

Regardless a bulk of papers devoted to the FMM, it is hard enough to find a detailed description of the effective C++ implementation of the FMM algorithms, which is platform-independent and scalable to cluster architecture. Moreover, as to our knowledge there is no available FMM subroutine for the standard packages for the molecular dynamic simulations, such as DL_POLY and AMBER. In this paper we basically follow the Greengard-Rokhlin algorithm [8, 7, 9] that was implemented by us in C++, and that we hope can be used for general applications.

The remainder of this paper is organized as follows. In *Section* 2. we remind the mathematical basis of the multipole decomposition of the $1/R$ potential into spherical harmonics. In *Section* 3. we describe the two-pass implementation of the Greengard and Rokhlin FMM in our code. *Section* 4. is devoted to the peculiarities of the calculation of the force field, which is the gradient of the potential and may have singularities at small angle differentiation. In *Conclusion* the results of the test runs of the FMM code are presented.

2. Mathematical Background

The Fast Multipole Method (FMM) is intended for the evaluation of $1/R$ potential fields in three dimensions. For a systems of N atoms (particles), located in a rectangular domain of size of size (a,b,c), the Coulomb sum

$$\Phi(r_j) = \sum_{i \neq j}^{N_{atoms}} \frac{q_i}{\|r_i - r_j\|} \tag{2}$$

is evaluated at the position of each j-th particle. Depending on particular problem, the potential field Φ may be considered as a field periodic in each direction, or as a non-periodic one. The force field used for the solution of the Newtonian equations of motion is then evaluated by taking the gradient of the potential 2:

$$F_i^{\alpha} = -q_i \nabla_{\alpha} \sum_{j \neq i} \frac{q_j}{\sqrt{(\mathbf{r}_i - \mathbf{r}_j)^2}} = q_i \sum_{j \neq i} \frac{q_j (r_i^{\alpha} - r_j^{\alpha})}{\left(\sqrt{(\mathbf{r}_i - \mathbf{r}_j)^2}\right)^3}. \tag{3}$$

The idea of the FMM is to substitute the summation over all pairs in 2, which requires $O(N^2)$ operations, by a direct sum over close neighbors with the interaction with far neighbors taken in the multipole approximation. For a large system of spatially distributed charged particles this can be effectively done by grouping the particles into rectangular blocks. These blocks are being sequentially coarse-graining, with the potential of the next coarse-grained block (*parent*) calculated as a sum of the previous level blocks (*children*), translated to the center of the parent according to the transformation laws of the multipole expansion.

Let us briefly recall the mathematical basis of multipole expansion. Let a charge of strength q be located at a point $P_0 = (x_0, y_0, z_0)$, which is sufficiently close to the center of coordinates $(0,0,0)$ in the sense that $r < a_0$, where $a_0 > 0$ and (ρ, θ, ϕ) are the spherical coordinates for P_0. The potential induced by the charge q in a point P outside the sphere of radius a_0 centered at $(0,0,0)$ can be evaluated using the decomposition with respect to the Legendre polynomials

$$\frac{1}{R} = \frac{1}{r\sqrt{1 - 2u\mu + \mu^2}} = \frac{1}{r} \sum_{n=0}^{\infty} P_n(u)\mu^n, \tag{4}$$

where $\mu = \rho / r$ is the ratio of the radius vector of the charge q to the radius vector of the observation point P. The angle between those two vectors is given by

$$u = \cos\gamma = \cos\theta\cos\alpha + \sin\theta\sin\alpha\cos(\phi - \beta), \tag{5}$$

where (ρ, α, β) are the spherical coordinates for the charge q located at P_0, and (r, θ, ϕ) are those of the observation point P.

The expansion 4 based on the Legendre polynomials $P_n(u)$ provides an effective way to calculate the potential induced by a charge located at P_0 at a distant point P. If a number of charges q_i are located near the center of coordinates $(0,0,0)$, the final equation for the potential they induce at a distant point P depends only on local coordinates of the observation point and the multipole moments

$$M_n^m = \sum_{i=1}^{k} q_i \rho_i^n Y_n^{-m}(\alpha_i, \beta_i), \tag{6}$$

where $(\rho_i, \alpha_i, \beta_i)$ are the spherical coordinates of the charges. The spherical functions and the Legendre polynomials are defined as

$$Y_n^m(\theta, \phi) \equiv \sqrt{\frac{(n-|m|)!}{(n+|m|)!}} P_n^{|m|}(\cos\theta) e^{\imath m \phi}, \tag{7}$$

where the associated Legendre polynomials are defined accordingly,

$$P_n^m(x) = (-1)^m (1-x^2)^{m/2} \frac{d^m}{dx^m} P_n(x).$$

For the negative values of m the spherical functions are calculated using the symmetry

$$Y_n^{-m}(\theta, \phi) = Y_n^{*m}(\theta, \phi). \tag{8}$$

The goal of the Fast Multipole Method (FMM) consists in achieving $O(N \log N)$ resource consumption by taking into account the Coulombic interaction with near neighbors only, taking the effect of other particles by multipole expansion. If the total number of particles is big enough, it is reasonable to calculate the multipole moments sequentially coarse-graining the moments into blocks: the larger the distance between the source charge and the observation point, the coarser is the block that comprises that charge.

To perform the calculation the volume occupied by the system of charged particles is subdivided into rectangular cells. For each cell the moments are evaluated according to 6. The potential induced by these charges outside the sphere of the radius a is then given by

$$\Phi = \sum_{n=0}^{\infty} \sum_{m=-n}^{n} \frac{M_n^m}{\rho^{n+1}} Y_n^m(\theta, \phi). \tag{9}$$

This is so-called multipole expansion of the electrostatic potential.

The relative error of the multipole expansion is given by

$$\left| \Phi(P) - \sum_{n=0}^{p} \sum_{m=-n}^{n} \frac{M_n^m}{\rho^{n+1}} Y_n^m(\theta, \phi) \right| \le \frac{A}{r-a} \left(\frac{a}{\rho} \right)^{p+1}, \tag{10}$$

where $A = \sum_{i=1}^{k} |q_i|$ is the sum of the modulus of all charges.

3. Implementation

3.1. Greengard and Rokhlin Algorithm

The idea of the FMM algorithm is to achieve $O(N \log N)$ order resource consumption by calculating the potential at each point in a two-pass algorithm. In the upward pass the multipole moments of the larger blocks are evaluated by translating the multipole expansion of each small block (child) to the center of the larger block (parent). In the downward pass the local expansion of the parent is converted into the local expansion of the children and is corrected by taking into account the effect of the *interaction list* – the blocks that are not direct neighbors, but stand next to the direct neighbors. The 3D picture of the interaction list of a given cell is shown in Fig. 1 below.

In our code we implemented the three-dimensional algorithm by Greengard and Rokhlin presented in [9]. Our scheme is not adaptive, but has the equal partition of the initial cell at the finest resolution level, given at the stage of object initialization. This may be not so effective for the inhomogeneous particle distribution, but saves the time for homogeneous particle distribution since the initialiazation of all objects is done only once at the beginning of calculations. However, if a given cell contains no particles at all it is ignored in calculations. For the fully adaptive implementation of the Greengard and Rokhlin FMM see [10].

In simulations we assumed N charged particles to be homogeneously distributed with the rectangular domain of size (a,b,c) in three dimensions. This is the cell of the most coarse grained level – we denote it as a zero level – and is a parent to 8 cells of the first level, the children, each of those has a half size $(a/2,b/2,c/2)$; the n-th resolution level contains 8^n cells, respectively. The number of the levels should be chosen so that each cells of the finest resolution level should contain a few charged particles. The Coulombic interaction of a given particle with all other particles in the same cell and all particles in the neighboring cells is calculated directly. The interaction with all other particles in neigboring cells is calculated by means of the multipole expansion of the potential.

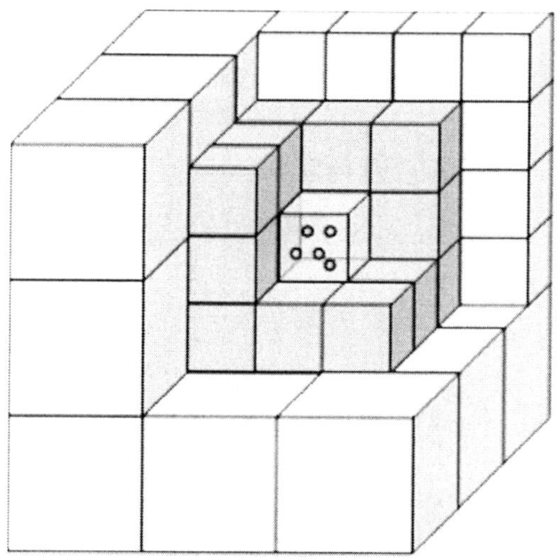

Figure 1. The location of neighbors and interaction list of a given cell containing a few particles in 3D space. The interaction list is indicated in white

Therefore all cells of the given level have the same size, and their parameters can be calculated only once, at the beginning of calculations. The uniform FMM works very well when all particles are uniformly distributed, i.e.,there are no big voids in the integration domain. If the particles are distributed not uniformly an adaptive method is required for effective calculations [11, 12].

To calculate the potential of the near particles that are not near neighbors, following [9], we use the concept of *interaction list*. The interaction list of a given cell, is a set of all cells of the same resolution level, those are the children of the neighbors of the parent of the given cell, but are not the near neighbors of the given cell at the same time. The effect of the distant particles, those are out of the interaction list is evaluated at the next resolution level. So, the structure of the Greengard-Rokhlin algorithm looks as follows:

1 Using the formula 6 we calculate the multipole moments of all particles in each cell of the finest resolution level. The moments are calculated with respect to the center of each cell.

2 **Upward pass**. To calculate the multipole moments of the cells of the next resolution level we sum up the multipole moments of all children of each cell by translating the multipole expansion of each child to the center of the parent, using the formula

$$^{l-1}M_j^k = \sum_{n=0}^{j} \sum_{m=-n}^{n} {}^{l}M_{j-n}^{k-m} \frac{\iota^{|k|-|m|-|k-m|} A_n^m A_{j-n}^{k-m} \rho^n Y_n^{-m}(\alpha,\beta)}{A_j^k}. \tag{11}$$

This process goes up to the zero level cell ($l=0$) – to the whole domain of calculations.

3 **Downward pass**. To calculate the electrostatic energy of a set of particles in a given cell it is convenient to consider the electrostatic potential, induced by distant particles in this cell, as a function of spherical coordinates centered in the middle of this cell. This is so-called local expansion:

$$\Phi = \sum_{n=0}^{\infty} \sum_{m=-n}^{n} L_n^m r^n Y_n^m(\theta,\phi). \tag{12}$$

At the *coarsest* level ($l=0$) there is no local expansion, since there are no charges external to the whole set: $\forall m,n\ L_n^m = 0$. On the next ($l=1$) level there are 8 children of the unique parent cell ($l=0$); all of them are neighbors, and hence there are no interaction lists at this level as well. The transformation of the multipole expansion to the local expansion starts from the next level ($l=2$), where the multipole moments of the previous level $^{l=1}M$ are converted to the local coefficients $^{l=2}L$:

$$^{l+1}L_j^k = \sum_{n=0}^{\infty} \sum_{m=-n}^{n} \frac{^{l}M_n^m \iota^{|k-m|-|k|-|m|} A_n^m A_j^k Y_{j+n}^{m-k}(\alpha,\beta)}{(-1)^n A_{j+n}^{m-k} \rho^{j+n+1}}, \tag{13}$$

where (ρ,α,β) are the coordinates of the parent with respect to the child. Going down, the potential in each i-th cell is calculated as a sum of the translation of the local expansion of its parent $p(i)$ and the multipole expansions of all cells in its interaction list $ilist(i)$.

The translation of the local expansion from the parent to the local expansion of the child is performed according to the translation formula:

$$^{child}L_j^k = \sum_{n=j}^{p} \sum_{m=-n}^{n} \frac{^{parent}L_n^m \iota^{|m|-|k|-|m-k|} A_{n-j}^{m-k} A_j^k Y_{n-j}^{m-k}(\alpha,\beta)}{(-1)^{n+j} A_n^m \rho^{j-n}}. \tag{14}$$

The contribution of all cells of the interaction list of a given cell is evaluated accordingly,

$$ {}^{l}_{s}L_j^k = \sum_{t \in ilist(s)} \sum_{n=0}^{\infty} \sum_{m=-n}^{n} \frac{{}^{l}_{t}M_n^m \, \iota^{|k-m|-|k|-|m|} A_n^m A_j^k Y_{j+n}^{m-k}(\alpha, \beta)}{(-1)^n A_{j+n}^{m-k} \rho^{j+n+1}}. \tag{15} $$

On the finest level of resolution the potential is evaluated at the position of each particle by means of local expansion

$$ \Phi(r, \theta, \phi) = \sum_{l=0}^{p} r^l \sum_{m=-l}^{l} L_m^l Y_l^m(\theta, \phi). \tag{16} $$

4 Calculate the contribution of particles in neighboring cells by direct calculation of Coulomb sums.

All formulas are given in accordance to the original paper [9], with the normalization constants

$$ A_n^m = \frac{(-1)^n}{\sqrt{(n-m)!(n+m)!}}. $$

3.2. Code Description

We have implemented the fast multipole algorithm by Greengard and Rokhlin [9] in an object-oriented C++ code. The modern compilers allow to link the C++ implemented objects together with the programs implemented in other programming languages, e.g.,FORTRAN, C, etc.. The main reasons for using the object-oriented C++ are the simplicity of the development of complex mathematical algorithms and minimization of errors at the stage of development. This results in a more optimized code, suitable for easy parallelization of calculations. In our implementation the Intel-8.1 compiler was used.

To perform the calculation of the electrostatic force field by means of fast multipole algorithm, we created a complex hierarchic object, called the **FMM**. The structures of the objects used in our code is shown in Fig. 2. The composed object FMM contains the data and all functions necessary for multipole expansion. The FMM is comprised of a number of primitive objects; their inheritance is shown in Fig. 2. At the stage of initialization of the FMM object the space domain of calculations, the maximal degree of polynomials (p, the order of expansion) and the number of layers (scales) of the space coarse-graining are specified. During the initialization all dynamical objects are allocated in memory, and all preliminary calculations needed for FMM methods are performed. Since the partition of the cell is the same for all levels, the directions from the center of the parent cell to the centers of its children are the same for all levels as well, but radius vectors from parent to children are related to the base ones by the scale factor 2^j only depending on the number of layer j. So, we have calculate all directed angles to the neighboring cells and the interaction list cells, and corresponding spherical functions for all cells of all layers, only once at the beginning of calculation. The results of these preliminary calculations are kept in the "**Store**" object, the FMM is inherited from.

Depending on the number of particles located in the calculation domain, the initial cell of size (a, b, c) is sequentially divided into a number sub-grids of half-size, called "**Layers** ".

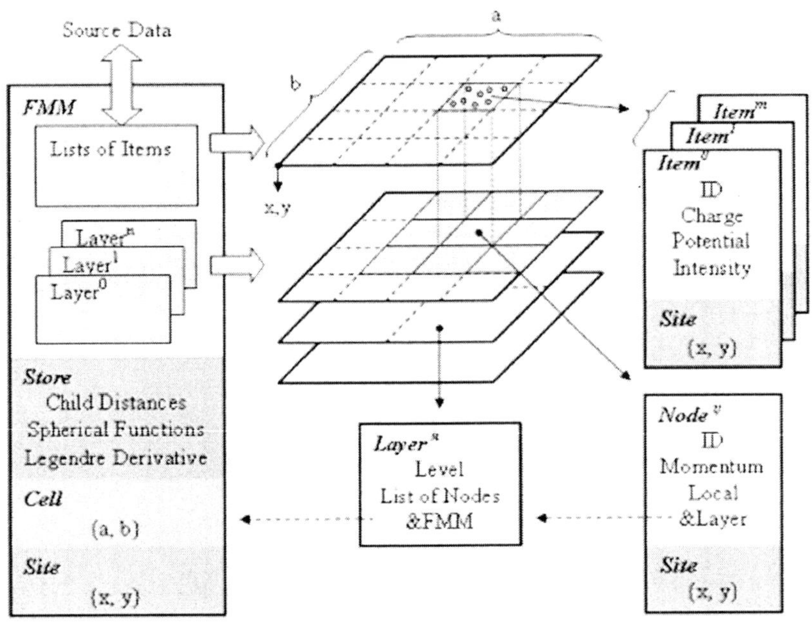

Figure 2. FMM object structure. For the sake of simplicity of imagination the 2D structures are drawn instead of their 3D counterparts implemented in C++ code

We do not use periodic boundary conditions in our code. The number of layers is to be chosen to balance the object initialization time and step time of the potential and the force field calculation. The coordinates and the charges of the particles are filled into the FMM object from source data. The FMM distributes the particles according to the cell of the finest resolution layer they belong to and forms the array of lists.

Having filled the lists of particles, the FMM performs the upward pass, calculating the moments ${}^{l}M$ for all layers of the hierarchy. On the downward pass the local moments ${}^{l}L$, which accumulate the effect of the interaction list to each cell, are calculated.

Since each charged particle is located in a certain cell of the finest coarse-graining level, the spherical coordinates of all charges are kept on this finest level only. All other levels contain the array of multipole M and local L expansion moments of the given order. The order of expansion is specified at the time of initialization of the FMM object. The arrays of the calculated M and L values are stored in the "**Node**" object related to each cell. The values of physical quantities – the potential, and the field intensity, i.e.,the gradient of the potential, are stored into the "**Item**" object, related to each cell.

To optimize the memory usage we use the symmetry of the spherical functions with respect to complex conjugation 8: we keep in memory only the L_l^m, M_l^m values for $m \leq 0$, the positive values of m were calculated using the symmetry.

4. Calculation of the Force Field

The exact value of the force acting on the i-th charge in its position $\mathbf{r}_i$ is equal to

$$\mathbf{F}_i = -q_i \nabla \sum_{j \neq i} \frac{q_j}{\sqrt{(\mathbf{r}_i - \mathbf{r}_j)^2}} = q_i \sum_{j \neq i} \frac{q_j(\mathbf{r}_i - \mathbf{r}_j)}{\left(\sqrt{(\mathbf{r}_i - \mathbf{r}_j)^2}\right)^3}, \tag{17}$$

or in the coordinate notation

$$F_i^\alpha = -q_i \nabla_\alpha \sum_{j \neq i} \frac{q_j}{\sqrt{(\mathbf{r}_i - \mathbf{r}_j)^2}} = q_i \sum_{j \neq i} \frac{q_j(r_i^\alpha - r_j^\alpha)}{\left(\sqrt{(\mathbf{r}_i - \mathbf{r}_j)^2}\right)^3}, \tag{18}$$

where $\alpha = x, y, z$ labels the Cartesian coordinates. To obtain the gradient of the potential Φ we first have to define the gradients 18 in spherical coordinates and then transform them to the Cartesian coordinated. Using the equation for local expansion 16, we calculate the gradient in spherical coordinates:

$$\begin{aligned} P_r &\equiv \frac{\partial \Phi}{\partial r} = \sum_{l=0}^{p} l r^{l-1} \sum_{m=-l}^{l} L_m^l Y_l^m(\theta, \phi) \\ &= \sum_{l=0}^{p} l r^{l-1} \sum_{m=-l}^{l} L_m^l Y_l^m(\theta, \phi), \\ P_\theta &\equiv \frac{1}{r}\frac{\partial \Phi}{\partial \theta} = \sum_{l=0}^{p} r^{l-1} \sum_{m=-l}^{l} L_m^l \frac{Y_l^m(\theta, \phi)}{\partial \theta}, \\ P_\phi &\equiv \frac{1}{r \sin\theta}\frac{\partial \Phi}{\partial \phi} = \frac{\iota}{\sin\theta} \sum_{l=0}^{p} r^{l-1} \sum_{m=-l}^{l} m L_m^l Y_l^m(\theta, \phi). \end{aligned} \tag{19, 20}$$

The force field is equal to the product of the charge by the strength of the field, $E = -\nabla\Phi$, so

$$F_i^\alpha = -q_i P_i^\alpha, \quad \text{where } \alpha = r, \theta, \phi. \tag{21}$$

The differentiation of the spherical functions with respect to cos θ, i.e., the differentiation of the associate Legendre polynomials with respect to x, is performed using the formula

$$\frac{\partial Y_l^m}{\partial \theta} = \frac{lx}{\sqrt{1-x^2}} Y_l^m - \sqrt{\frac{(l+m)(l-m)}{1-x^2}} Y_{l-1}^m, \quad x = \cos\theta, \tag{22}$$

where all $Y_l^m, m > 1$ are assumed to be 0.

The singularity at the zero angle ($x \to 1$) in the latter equation is tackled using the sin θ term in the equations for P_r, P_θ, P_ϕ. To do that, following [13], we use the following relations

$$\begin{aligned} \frac{\partial}{\partial \theta} P_n^m(\cos\theta) &= \frac{\partial}{\partial \theta}\left[(-1)^m \sin^m \theta \left(\frac{\partial}{\partial \cos\theta}\right)^m P_n(x)\right] \\ &= \frac{m \cos\theta}{\sin\theta} P_n^m(\cos\theta) + P_n^{m+1}(\cos\theta), \end{aligned}$$

from there, for the spherical functions with $m = \pm 1$, we get

$$\frac{\partial}{\partial \theta} Y_n^{\pm 1} = -\frac{1}{\sqrt{n(n+1)}} P_n'(1) e^{\pm \iota \phi}. \tag{23}$$

The derivatives of the Legendre polynomial are calculated recursively:

$$P_n'(1) = P_{n-1}'(1) + nP_n(1). \tag{24}$$

In a similar way we treat the azimutal component P_ϕ, for that we have

$$\frac{1}{\sin\theta} Y_n^{\pm 1} = -\frac{1}{\sqrt{n(n+1)}} P_n'(1) e^{\pm \imath\phi}. \tag{25}$$

The transformation of a vector field $\vec{P}$ into Cartesian system is:

$$P_x = P_r \sin\theta\cos\phi + P_\theta \cos\theta\cos\phi - P_\phi \sin\phi, \tag{26}$$

$$P_y = P_r \sin\theta\sin\phi + P_\theta \cos\theta\sin\phi + P_\phi \cos\phi, \tag{27}$$

$$P_z = P_r \cos\theta - P_\theta \sin\theta. \tag{28}$$

5. Results and Conclusions

We have tested the code to calculate force field for the number of particles from 10^3 to 10^6, homogeneously distributed in the unit domain (1 .0, 1.0, 1.0). The number of spherical harmonics was taken in the range from 3 to 9 for different number of layers. The resulting time cost includes both the time spent for the initialization of the 3DCell objects and the calculation time.

We used neither rotations nor plain waves decomposition for the speed-up of the $L \to L$ and $M \to L$ transformations. The reason is the rotation itself leads to some accuracy loss, and we just prefer to keep small number of harmonics instead. The time cost of calculations of the force field are presented in Fig. 3.

The accuracy of the method was estimated by calculating the RMS relative error of the potential evaluation and its maximal error. The RMS error was evaluated as

$$\sigma = \sqrt{\frac{1}{N}\sum_{i=1}^{N} \varepsilon^2}, \quad \text{where } \varepsilon = \frac{\Phi_i - \Phi_i^{exact}}{\Phi_i^{exact}}. \tag{29}$$

The maximal error was evaluated as the maximum absolute value of ε. The exact value of the potential Φ_i^{exact} was calculated at the location of each particle by means of direct summation. The dependence of the RMS relative error and the maximal relative error on the order of the multipole expansion is presented in Fig. 4.

The evaluation of the force field, which includes the derivatives of the Legendre polynomials (24,25) contributes not more than one order to the relative error. Therefore, our experiments with the relative error of the FMM method show that for practical needs of molecular dynamic calculations it is enough to keep the 4th or 5th degree of the multipole decomposition to provide for the 10^{-4} or higher accuracy of the force field calculations. For this reason we did not perform the rotation optimization of the 13 and 14 transformations: for $p = 4, 5$ the computational burden of order p^4 is still bearable.

All tests were performed on Pentium4 2.4 GHz processor under Red Hat Linux 9.2 operation system. The Intel-8.1 C++ compiler was used for the compilation of the code.

The memory allocation in MBytes depending on the number of layers and the order of polynomial expansion is shown in Fig. 5.

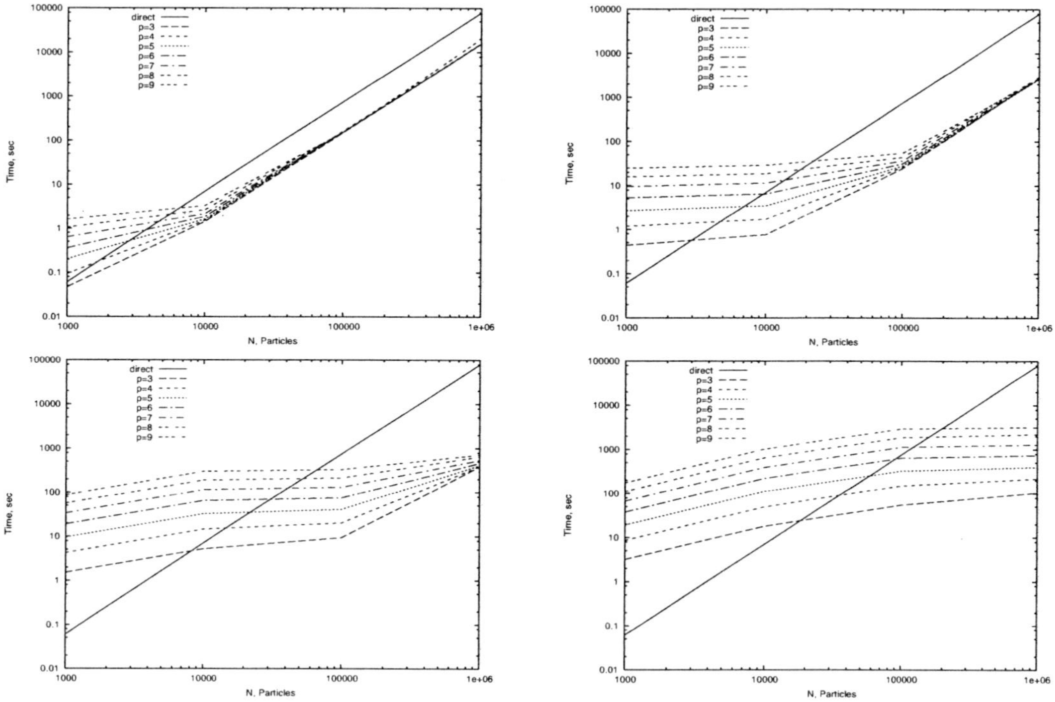

Figure 3. Timings of calculation the force field for the systems of $N = 10^3, 10^4, 10^5, 10^6$ particles. Calculated for the coarse-graining of the initial cell into 3, 4, 5 and 6 layers, respectively. Different types of the dashed lines in the graphs show different orders of multipole expansion $p = \overline{3,9}$. The time cost of direct calculation $O(N^2)$ is shown by the solid line in each of the four graphs

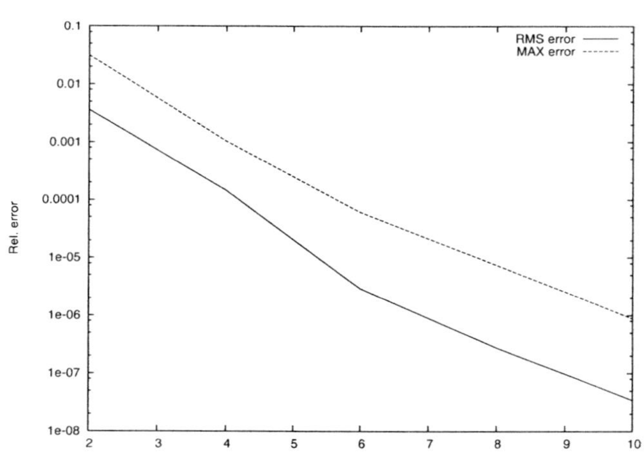

Figure 4. The dependence of the RMS relative error and the maximal relative error on the order of the multipole expansion for the Coulombic potential

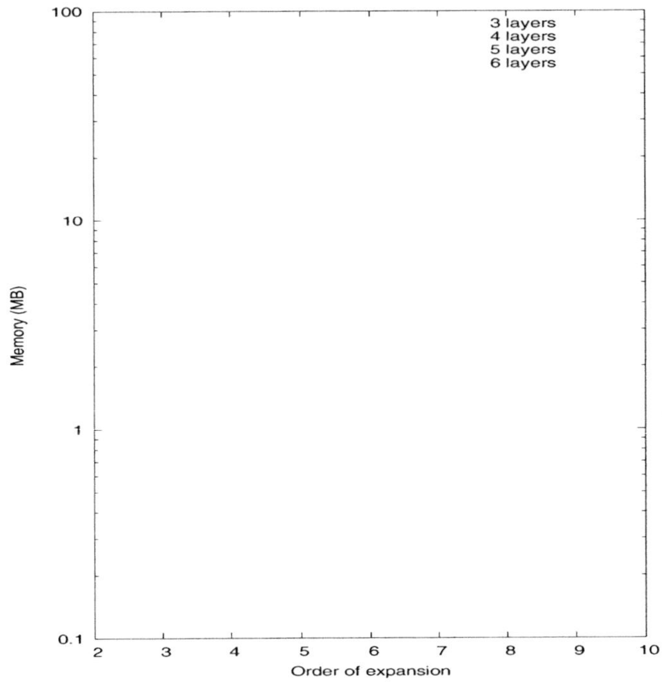

Figure 5. Memory allocation in MBytes as a function of the order of polynomial expansion (p) shown for different number of layers

References

[1] A.J. Chorin, *J. Fluid. Mech.*, (1973), 57, 785-796.

[2] P. Ewald, *Ann. Phys.*, (1921), 79,253.

[3] T. Narumi, R. Susukita, H. Furusawa, K. Yasuoka, A. Kawai, T. Koishi, T. Ebisuzaki, MDM version of AMBER, 2000.

[4] N. Okimoto, K. Yamanaka, A. Suenaga, Y. Hirano, N. Futatsugi, T. Narumi, K. Yasuoka, R. Susukita, T. Koishi, H. Furusawa, A. Kawai, M. Hata, T. Hoshino, T. Ebisuzaki, *Chem-Bio Informatics J.*, (2003), 3(1), 1-11.

[5] J.E. Barnes and P. Hut, *Nature*, (1996), 324(4), 446-449.

[6] A.A. Appel, *SIAM J. Sci. and Statist. Computing*, (1985), 6, 85-93.

[7] L. Greengard and V. Rokhlin, *J. Comp. Phys.*, (1987), 73, 325-348.

[8] L. Greengard, *The rapid evaluation of potential fields in Particle Systems* , ACM Press, 1987.

[9] L. Greengard and V. Rokhlin, *Acta Numerica*, (1997), 229-269.

[10] H. Cheng, L. Greengard, and V. Rokhlin, *J. Comp. Phys.*, (1999), 155, 468-498.

[11] J.P. Singh, C. Holt, T. Totsuka, A. Gupta, and J. Hennessy, *J. Par. Distr. Comp.*, (1995), 27, 118-141.

[12] H. Schwichtenberg, H. Winter, and H. Wallmeier, *Parallel Computing*, (1999), 25, 535-546.

[13] K. Esselink, *Comp. Phys. Comm.* (1995), 8, 375-395.

In: Molecular Simulation Studies in Material ... ISBN 1-59454-607-X
Editor: Kholmirzo Kholmurodov, pp. 161-172

Chapter 13

SEQUENCING ANALYSIS OF MUTANT ALLELE *CDC28-srm* OF PROTEIN KINASE *CDC28* AND MOLECULAR DYNAMICS STUDY OF GLYCINE-RICH LOOP IN WILD TYPE AND MUTANT ALLELE *G16S* OF *CDK2* AS MODEL

N. A. Koltovaya,* D. A. Kretov and Kh. T. Kholmurodov
Department of Radiation and Radiobiological Research
Joint Institute for Nuclear Research
141980 Dubna, Moscow Region, Russia
A. S. Guerasimova
Max-Planck-Institute for Molecular Genetics
14195 Berlin, Germany

Abstract

The central role that cyclin-dependent kinases play in the timing of cell division and the high incidence of genetic alteration of CDKs or deregulation of CDK inhibitors in a number of cancers make CDC28 of yeast *Saccharomyces cerevisiae* very attractive model for studies of mechanisms of CDK regulation. We have found that certain gene mutations including *cdc28-srm* affect cell cycle progression, maintenance of different genetic structures [1] and increased cell sensitivity to ionizing radiation [2]. A *cdc28-srm* mutation is not temperature-sensitive mutation and differs from known *cdc28-ts* mutations because it has the evident phenotypic manifestations at 30 °C. Sequencing analysis of *cdc28-srm* revealed a single nucleotide substitution G20S. This is a third glycine in a conserved sequence GxGxxG in the G-rich loop positioned opposite the activation T-loop. Despite its demonstrated importance, the role of the G-loop has remained unclear. The crystal structure of the human CDK2 has served as a model for the catalytic core of other CDKs, including CDC28. Nanoseconds long molecular dynamics (MD) trajectories of the CDK2/ATP complex were analyzed. The MD simulations of substitution CDK2-G16S (CDC28-G20S) shows a conformational change of CDK2 structure resulting in the moving of the G-loop away from ATP and a new rearrangement of amino acids in the T-loop.

*E-mail address: koltovaya@jinr.ru

1. Introduction

The cyclin dependent kinases (CDKs) are a subfamily of serine/threonine-specific protein kinases. The enzymes catalyze a transfer of the gamma-phosphate of adenosine triphosphate (ATP) to a protein substrate. CDKs are crucial regulators of the timing and coordination of eukaryotic cell cycle events. Transient activation of these kinases at specific cell stages is thought to trigger the principal transitions of the cell cycle, including DNA replication and entry into mitosis. In yeast, both transitions are controlled by single CDK (CDC28 in *Saccharomyces cerevisiae* [3]). In human cells, cell cycle events are governed by several CDKs. CDK4-cyclin D is necessary for passage through G1, CDK2-cyclin E is necessary for the transition from G1 to S phase, CDK2-cyclin A is necessary for progression through S, and CDC2-cyclin B is necessary for the transition from G2 to M phase. Two of these proteins CDK2 and CDC2/CDK1 have long been studied extensively. Both are closely related to yeast CDC28/CDK1 (CDK2 62% and CDC2 60% identity in amino-acid sequence).

The central role that cyclin-dependent kinases play in the timing of cell division and the high incidence of genetic alteration of CDKs or deregulation of CDK inhibitors in a number of cancers make CDC28 very attractive as a model for studies of mechanisms of CDK regulation. We have found that certain gene mutations in *Saccharomyces cerevisiae* including *cdc28-srm* affect cell cycle progression and maintenance of different genetic structures [1]. The *cdc28-srm* was accompanied by decreased mitotic stability of excess natural chromosomes and recombinant circular plasmids containing centromeres as well as decreased rates of spontaneous mitochondrial rho$^-$ mutations [1], increased cell sensitivity to ionizing radiation [2] and confered checkpoint defects [4]. A *cdc28-srm* mutation is not temperature-sensitive mutation and differs from known *cdc28-ts* mutations because it has the evident phenotypic manifestations at 30°C.

Crystallographic studies of several eukaryotic protein kinases have shown that they all share the same fold and tertiary structure. The crystal structure of the human CDK2 [5],[6] has served as a model for the catalytic core of other CDKs, including CDC28. The CDK2 structure is bilobed with an N-terminus that is primarily beta-sheet and a C-terminus that is primarily alpha-helix. ATP binds in a cleft between the two lobes. CDKs are inactive as monomers. Cell cycle dependent oscillations in CDK activity are induced by complex mechanisms that include binding to positive regulatory subunits (cyclins) and phosphorylation at positive (pT160 in T-loop) and negative (pT14, pY15 in G- loop) regulatory sites.

The aim of the present work is to determine the nature of *cdc28-srm* mutation and to simulate the structure of the mutant allele through the experimental and molecular dynamics simulation studies.

2. Materials and Methods

2.1. Preparing of Chromosomal DNA

Preparing of chromosomal DNA from yeast cells *cdc28-srm* was done by the standard technique [7].

2.2. Analysis of Nucleotide Sequence of Mutant Allele cdc28-srm

A set of coloured fluorescent terminators (ABIPRISM Dye Terminator Kit with AmpliTaq, Perkin-Elmer) and ORF specific primers were employed for determining the nucleotide sequence of the mutant allele *cdc28-srm* in an ABIPRISM 3700 sequencer according to standard procedure.

2.3. MD Simulations of Conformational Changes of Proteins

For the MD simulations the SANDER modules of the program package AMBER 8.0 [8] and of the modified version of AMBER 7.0 for a special-purpose computer MDGRAPE-2 [9] were used. The starting geometries for the simulations were prepared using X-ray structures from the Prookhaven Protein Data Bank ( ). The all-atom force field [10] was used in the MD simulations. A system was solvated with TIP3P molecules [11] generated in a spherical (non-periodic) water bath. The system temperature was kept constant by the Berendsen algorithm with coupling time of 0.2 ps [12]. Only bond lengths involving hydrogen atoms were constrained using the SHAKE method [13]. The integration time step in the MD simulations was 1 fs. The simulation procedures were the same in all calculations [14]. Firstly, for each system a potential energy minimization was performed on an initial state. Next, the MD simulation was performed on the energy-minimized states. The temperatures of the considered systems were gradually heated to 300 K for and then kept at 300 K for the next 1 million time steps [15],[16]. The trajectories at 300 K for 2-ns were compared and studied in detail. The result of simulations and images of simulated proteins were analyzed by RasMol [17] and MOLMOL [18] packages.

3. Results

3.1. cdc28-srm is a Substitution of Glycine with Serine in Position 20

Sequencing analysis of *cdc28-srm* revealed a single nucleotide substitution of glycine with serine in position 20 (G20S). This is a third glycine in a conserved sequence GxGxxG in G-rich loop. Glycine is the smallest amino acid, and serine is larger. Both amino acids belong to a neutral polar class. They are well solvable in water because their polar R-groups may form hydrogen bonds with water (for glycine and serine polar R-groups – H- and HO-CH2-, respectively). It is worth noting that the observation of a key mutation in a protein extended structure is important and it allows understanding the mechanism of the protein kinase regulation activity. In our case we have obtained a mutation in a highly conservative sequence in G-rich loop. Despite its demonstrated importance, the role of G-loop has remained unclear. We will simulate the consequence of the conserved glycine substitution in the G-rich loop on the structure of the protein kinase. We use the human CDK2 as model. Substitution of G20S in yeast CDC28 corresponds to G16S in human CDK2.

At the beginning we analyze inactive complex CDK2/ATP. Analysis of the CDK2/ATP binary complex [5] indicates that ATP interacts with several residues lining the cleft between the two lobes. The adenine base is positioned in a hydrophobic pocket between the beta-sheet of the small lobe and the L7 loop between beta-5 and alpha-2. The ATP phosphates are held in position by ionic and hydrogen-bonding interactions with several residues, including K33, D145, and the backbone amides of the G-rich loop between beta-1 and beta-2.

3.2. The CDK2/ATP Structural Conformations

We simulated the CDK2-G16 /ATP of the wild-type structure and compared the conformational changes with CDK2-S16/ATP structure. The calculation results of the wild-type CDK2-G16/ATP and CDK2-S16/ATP structural conformations are shown in Figs. 1 and 2. The left pictures display the initial states, the right pictures – the final 1-ns states, respectively. For CDK2-G16 and CDK2-S16 structures we have also generated a number of the animation movies to display the real-time dynamical motions. We concentrate on the positional changes between the ATP, residue 16 and T-loop (the latter covers a left bottom alpha-helix shown in Fig. 1).

If to make the comparison of the pictures of the initial and final states of wild-type CDK2-G16/ATP structure, it is easy to see that there is not a visually large difference between these two structures. Moreover, the position of the amino acid residue 16 does not change within 1 million time step MD steps in comparison to that at the initial configuration. So, for the wild-type protein we found that it keeps its conformation stable relatively to the original state.

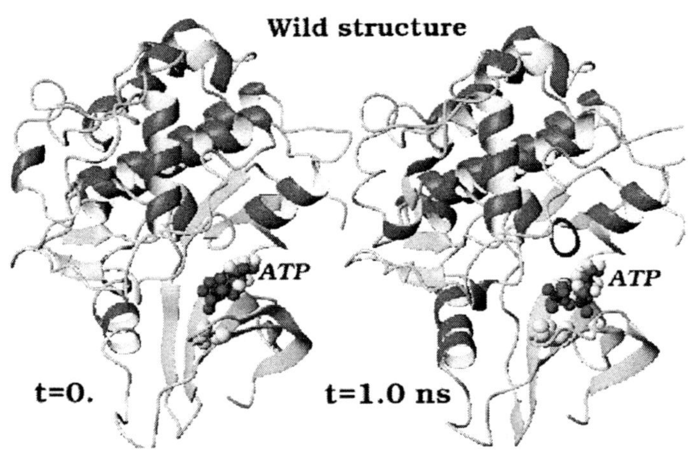

Figure 1. The initail and final (1-ns state) structures of the CDK2/ATP of the wild-type complex. The ATP molecule and residue 16 of the G-loop are drawn by balls.

Regarding the CDK2-S16 variant (Fig. 2), a completely different picture of the dynamics and conformational changes has seen. First of all, the amino acid residue 16 moves dramatically far from the ATP location site. In comparison to the wild structure, the distance between the position S16 and ATP, as it seen from the snapshots, increases of about 2-2.5 times at average. At the same time, such movement results on a relative shift of the T160 residue and the whole T- and G-loops positions.

On the one hand, it is obvious that the increase of the ATP-S16 distance influences on the picture of the hydrogen bonds formation involving the ATP and G-loop. On the other

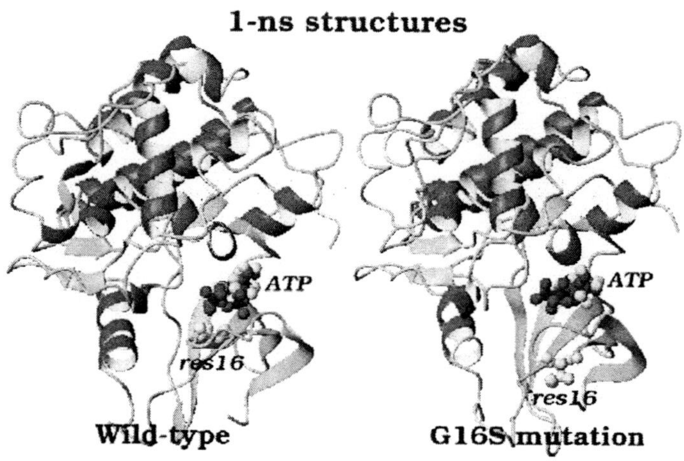

Figure 2. The comparison of the final (1-ns state) structures are shown (left: wild-type, right: S16 protein). The ATP molecule and the residue 16 of the G-loop are drawn by balls.

hand, the mutation induced CDK2' conformational change results in an interhelical protein movement, covering a phosphorylation point (viz. T160).

3.3. The G16S Mutation Induced Corruption of a Hydrogen Bond Network

In Fig. 3 we display the T160, ATP and S16 positions (an "activation triangle") for the final (1-ns) state to estimate (though indirectly) a possibility of the hydrogen bond formation in the ATP and G-loop region. We note that in general a hydrogen bond is regarded as being effective when the distance between the hydrogen atom of the proton donor and the proton acceptor is less then 2.6 - 2.4 Å. For the above "activation triangle" the calculated values of the distances between the ATP-res16 and ATP-T160 are shown in Figs. 4-5, respectively.

From Fig. 4 we observe that the ATP-res16 distance for the wild-type and mutant structures has a completely different behavior (Fig. 4). It is easily seen from the Fig. 4 that the ATP-res16 distance for the wild-type structure lies within $\sim$2.5 Å during the all 1-ns dynamical changes. At the same time for the mutation structure, ATP-res16 distance lies within of 7.5 Å. That is the G16S mutation causes the increase of the ATP-res16 distance of about three times large then that of the wild-type structure. Thus, the mutation induced conformational changes in the CDK2 structure are obviously corrupt the all hydrogen bond network in the ATP-res16 binding site.

In Fig. 5 we present the time dependences dynamics for the ATP-T160 distance. We see that the ATP-T160 distance for the wild-type protein is comparably shorter then that for the CDK2-S16/ATP variant. The behavior of the ATP-T160 distance agrees with the results shown in the previous figure.

Figure 3. The positions of the T160, ATP and res16 (an "activation triangle") are shown. The ATP molecule, residues T160 and 16 are drawn by balls.

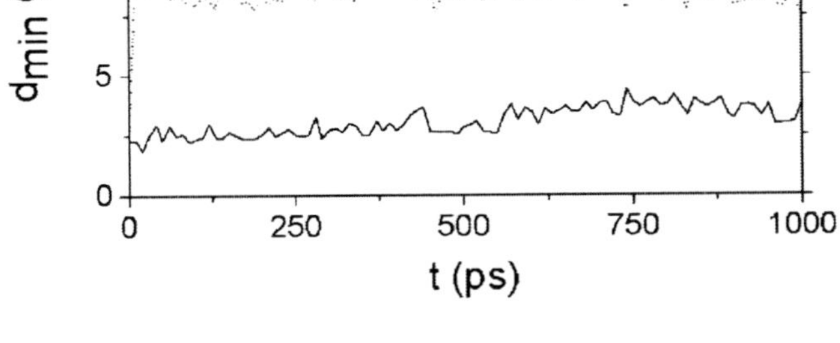

Figure 4. The time dependences of the ATP-res16 distance are shown for the CDK2-G16/ATP (solid line) and CDK2-S16/ATP (dot line) in accordance to the "activation triangle".

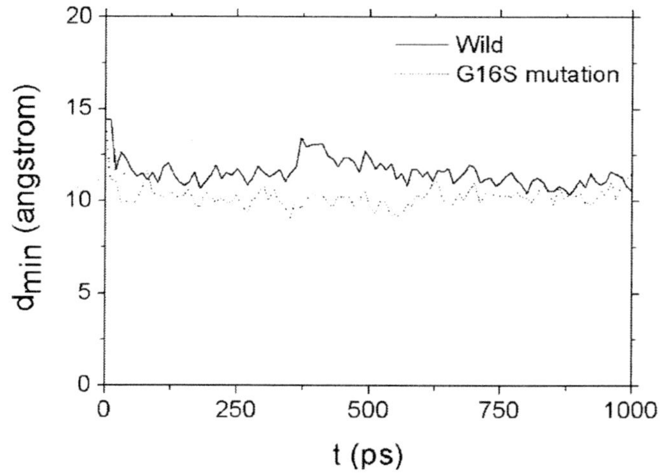

Figure 5. The time dependences of the ATP-T160 distance are shown for the CDK2-G16/ATP (solid line) and CDK2-S16/ATP (dot line) in accordance to the "activation triangle".

3.4. Rotational Changes of the Amino Acid Residues Around of Phosphorilated Regulatory Site T160

The CDK2/ATP' dynamical peculiarities around of the posphorilation site (T160) we have analyzed in detail, displaying by the snapshots and animation movies all amino acid positions in the T-loop.

From these data we have catches a rotation of the two amino acids neighboring T160 (viz. residues 159 and 161) that possesses different behavior. The residue 159 (Y159) is tyrosine with a non-charged polar R-group, and the residue 161 (H161) is histydine with a polar charged R-group. A summary of these observations is presented through the following two set of snapshots (Figs. 6 and 7).

From Fig. 6 (left snapshot: initial state) we see that two carbon rings of residues Y159 and H161 are occupy a well-separated positions. The carbon rings during their dynamical motions abruptly exchange (at around of 0.5 ns from the start) their orientation positions. The rings have begun to "screen" the amino acid residue T160. Such "screening" does not change until the end of the dynamics (right snapshot: final 1-ns state). The overlapping of the carbon rings, on the contrary, does not manifest itself for the S16 variant (Fig. 7). During the all simulation period there has to be no "screening" of the site T160 to seen.

Thus, the mutation induced changes for the protein structure affect the overlapping of the carbon rings at a key phosphorylation point T160. The "screening" phenomena described above we connect with the positional changes of the activation T-loop induced by the mutation at the residue 16.

From the "activation triangle" described above we have estimated the T160-res16 distance (see Fig. 8) for the wild-type CDK2/ATP, CDK2-S16/ATP and wild-type monomeric (without ATP) CDK2 proteins. It is seen from Fig. 8 that the T160-res16 distance in the mutant CDK2-S16/ATP structure is significantly larger then those of the wild-type CDK2-

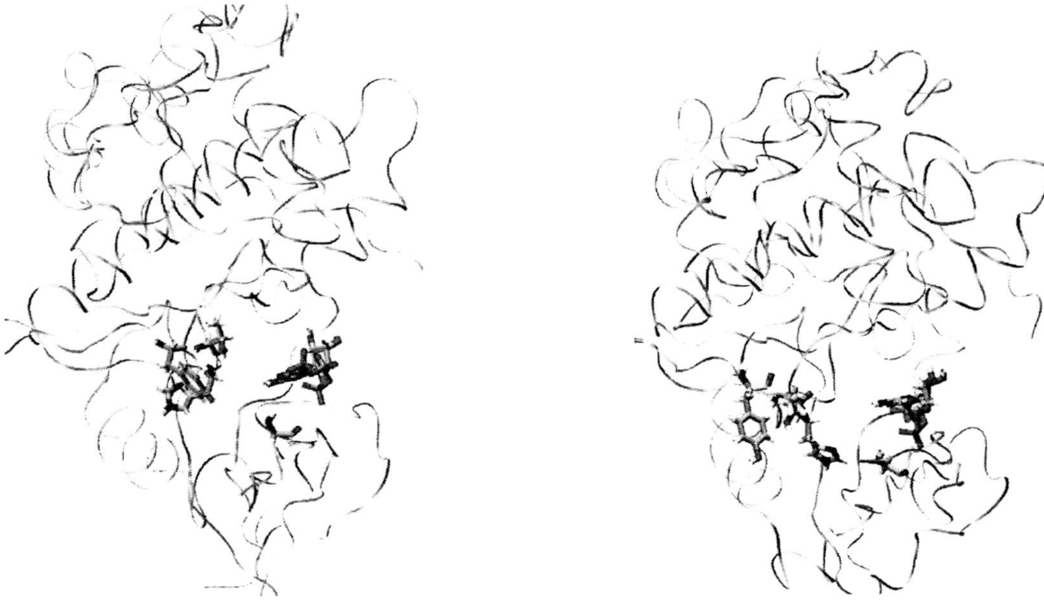

Figure 6. The initial (left) and final (right) configurations of the wild-type protein. The pictures display the orientational changes of two neighboring amino acids around of phosphorylation site (viz. the residue T160). It is seen that the Y159 and H161 exchange their relative orientations with respect to the residue T160.

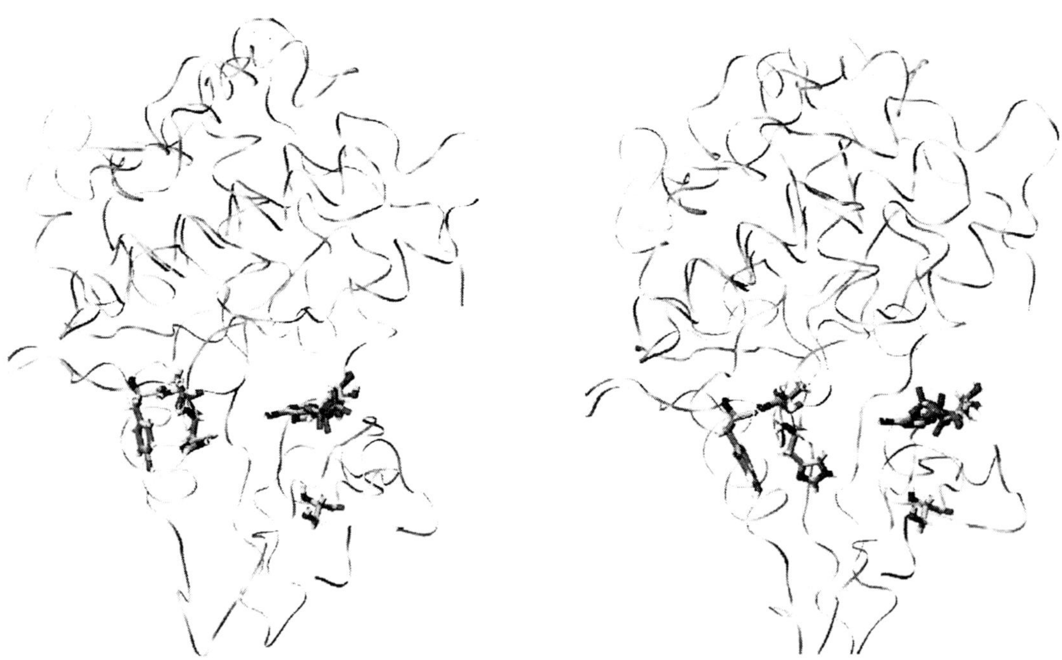

Figure 7. The initial (left) and final (right) configurations of the CDK2-S16 protein. The snapshots display the orientational changes of two neighboring amino acids (Y159 and H161) around of a phosphorylation site (T160). The positions of the ATP molecule and the res16 are also drawn.

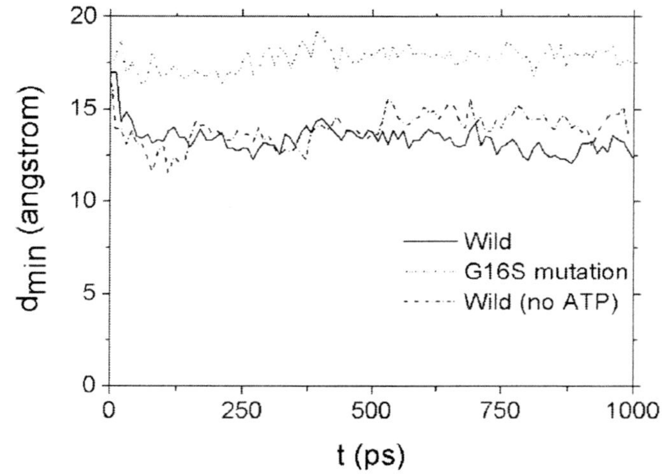

Figure 8. The time dependences of the T160-res16 distance are shown for the CDK2-G16/ATP wild-type (solid line), a wild-type CDK2 (dashed line) and CDK2-S16/ATP (dot line) in accordance to the "activation triangle.

G16/ATP ones. At the same time, the T160-res16 distance for the wild-type CDK2/ATP and wild-type monomeric CDK2 structures follows almost the same law. This fact probably means that the above "screening" phenomena do not origins from the presence of ATP molecule or T160-ATP exchanges.

4. Discussion

The G-loop enables protein kinase to adopt a wide range of backbone conformations. The significance of this domain is demonstrated by the fact that substitution of the glycine residues in the G-loop, particularly the first and the second glycine (GxGxxG) with either alanine or serine results in a dramatic decrease in cAPK activity. The functional importance of the G-loop has been described in detail for cAPK [19]-[22], but its importance for CDK regulation has not been yet discussed. It is believed that the G-loop catalytic function – that is, correct ATP binding and alignment – is the same as its function in cAPK, but it exhibits a new inhibitory function for CDK [23]. We have obtained the mutation *cdc28-srm* that posseses a pleotropic manifestation in yeast cells. Analyses have shown that this mutation is localized in the glycine-rich loop (G-loop) and is a substitution of the third glycine with serine.

The crystal structure of inactive monomeric CDK2 showed that the T-loop (residues 147-153) would block access of substrates to the active site and that ATP would bind with the wrong geometry for efficient catalysis. Binding to cyclin A simultaneously moves the T-loop of CDK2 away from the substrate binding cleft and repositions of the G-loop (residues 11-18), so that they can interact properly with the ATP phosphates. This complex has a low but detectable activity. Less dramatic changes occur in the structure of the CDK2/cyclin A complex following activation by phosphorylation of T160 [24]. The phosphorylation of

T160 might relieve a proposed steric interference between the substrate and the unphosphorylated T-loop and help to organize an acidic patch containing E162, E208 and D235 by repositioning E162, a T-loop residue. The serine of the peptide substrate is hydrogen bonded to the ATP gamma-phosphate oxygen, to the catalytic aspartate D127, and to the conserved lysine K129 [25].

MD simulation analysis shows that in CDK2-S16/ATP the distance between ATP and the residue 16 in the G-loop dramatically increased. The shift is equal 5 Å. It's known that the G-loop was changed during the early stage of inactive (CDK2/ATP), partly active (CDK2/cyclin A/ATP), and fully active (pT160-CDK2/cyclin A/ATP) CDK2 simulations in comparison with its conformation as found in the crystal structures [23]. The G-loop moves away from the ATP phosphate moiety binding site after the interaction of CDK2 with cyclin A and again after CDK2/cyclin A/ATP complex phosphorylation at the T160 site. The shift of the G-loop is equal to 3.5 Å (CDK2/cyclin A/ATP) and 8.6 Å (pT160-CDK2/cyclin A/ATP) in comparison with the G-loop position found in the CDK2/ATP system. It's interesting that we observe this shift in the mutant allele CDK2-G16S/ATP as well.

We find unexpected consequence of substitution G16S – new orientation of two neighboring amino acids of T160 in the T-loop. It may influence on the interactions of the protein kinase with cyclins. In further studies we intent to examine several other complexes of the CDK2, including cyclin A and substrate.

References

[1] A.B. Devin, T. Prosvirova, V.T. Peshekhonov, O.V. Chepurnaya, M.E. Smirnova, N.A. Koltovaya, et al, The Start gene *CDC28* and genetic stability of yeast, *Yeast*, (1990), 6, 231-243.

[2] N.A. Koltovaya, I.P. Arman, A.B. Devin, Mutations of the *CDC28* gene and the radiation sensitivity of *Saccharomyces cerevisiae*, *Yeast*, (1998), 14, 133-146.

[3] M.D. Mendenhall, A.E. Hodge, Regulation of Cdc28 cyclin-dependent protein kinase activity during the cell cycle of the yeast, *Microbiol. Mol. Biol. Rev.*, (1998), 62, 1191-1243.

[4] N.A. Koltovaya, K.Yu. Kadyshevskaya, A.B. Devin, DNA repair and cell cycle regulation: CDC28 gene and new mitotic checkpoint genes, *Yeast*, (1995), 11, 73.

[5] H.L. De Bondt, J. Rosenblah, J. Jancarik, H.D. Jones, D.O. Morgan, S.-H. Kim, Crystal structure of cyclin-dependent kinase 2, *Nature*, (1993), 363, 595-602.

[6] P.D. Jeffrey, A.A. Russo, K. Polyak, E. Gibbs, J. Hurwitz, J. Massague, N.P. Paoletich, Mechanism of CDK activation revealed by the structure of a cyclin A – CDK2 complex, *Nature*, (1995), 373, 313-320.

[7] E.D. Green, M.V. Olson, Systematic screening of yeast artificial-chromosome libraries by use of the polymerase chain reaction, *PNAS*, (1990), 87, 1213-1217.

[8] D.A. Case, D.A. Pearlman, J.W. Caldwell, T.E. Cheatham III, W.S. Ross, C.L. Simmerling, T.A. Darden, K.M. Merz, R.V. Stanton, A.L. Cheng, J.J. Vincent, M. Crowley, D.M. Ferguson, R.J. Radmer, G.L. Seibel, U.C. Singh, P.K. Weiner, P.A. Kollman, AMBER 8.0, University of California, 2003.

[9] N. Okimoto, K. Yamanaka, A. Suenaga, Y. Hirano, N. Futatsugi, T. Narumi, K. Yasuoka, R. Susukita, T. Koishi, H. Furusawa, A. Kawai, M. Hata, T. Hoshino, T. Ebisuzaki, Molecular dynamics simulations of prion proteins - effect of Ala117 $\Rightarrow$ val. mutation, *Chem-Bio Informatics J.*, (2003), 3(1), 1-11; T. Narumi, R. Susukita, H. Furusawa, K. Yasuoka, A. Kawai, T. Koishi, T. Ebisuzaki, MDM version of AMBER, 2000.

[10] W.D. Cornell, P. Cieplak, C.I. Bayly, I.R. Gould, Jr.K.M. Merz, D.M. Ferguson, D.C. Spellmeyer, T. Fox, J.W. Caldwell, P.A. Kollman, *J. Am. Chem. Soc.*, (1995), 117, 5179-5197.

[11] W.L. Jorgensen, J. Chandrasekhar, J.D. Madura, Comparison of simple potential functions for simulating liquid water, *J. Chem. Phys.*, (1983), 79, 926-935.

[12] H.J.C. Berendsen, J.P.M. Postma, W.F. van Gunsteren, A. DiNola, J.R. Haak, Molecular dynamics with coupling to an external bath, *J. Chem. Phys.*, (1984), 81, 3684-3690.

[13] J.P. Ryckaert, G. Ciccotti, H.J.C. Berendsen, Numerical-integration of cartesian equations of motion of a system with constraints - molecular-dynamics of N-alkanes, *J. Comput. Phys.*, (1997), 23, 327-341.

[14] Kh. Kholmurodov, Molecular-dynamics simulations of rhodopsin and prion proteins: the effect of disease-related amino acid mutations on their structural conformations, *PEPAN (Physics of Particles and Nuclei)*, (2005), 36(2), 1-16; Y. Hirano, E. Ebisuzaki, MD simulations on the influence of disease-related amino acid exchanges in the human prion proteins, *Chem-Bio Informatics J.*, (2003), 3(2), 86-95.

[15] Kh. Kholmurodov, T. Ebisuzaki, MD simulation on the structural changes and conformational dynamics of the retinal proteins (rhodopsins) in and explicit solvent, *ICMS-CSW2004*, (2004), C4, 9-11.

[16] Kh.T. Kholmurodov, et al., Methods of molecular dynamics for simulations of the physical and biological processes, *PEPAN (Physics of Elementary Particles and Atomic Nuclei)*, (2003), 34(2), 474-501.

[17] R.A. Sayle, E.J. Milner-White, RasMol: Biomolecular graphics for all, *Trends in Biochem. Sci.*, (1995), 20, 374-376.

[18] R. Koradi, M. Billeter, K. Wuthrich, MOLMOL: a program for display and analysis of macromolecular structure, *J. Mol. Graphics*, (1996), 4, 51-55.

[19] W. Hemmer, M. Mc Glone, I. Tsigelny, S.S. Taylor, Role of the glycine triad in the ATP-binding site of cAMP-dependent protein kinase, *J. Biol. Chem.*, (1997), 272, 16946-16954.

[20] I. Tsigelny, J.P. Greenberg, S. Cox, W.L. Nichols, S.S. Taylor, L.F. Ten Eyck, 600 ps molecular dynamics reveals stable substructures and flexible hinge points in cAMP dependent protein kinase, *Biopolymers*, (1999), 50, 513-524. 18.0e-6

[21] R.T. Aimes, W. Hemmer, S.S. Taylor, Serine-53 at the tip of the glycine-rich loop of cAMP-dependent protein kinase: Role in catalysis, P-site specificity, and interaction with inhibitors, *Biochemistry*, (2000), 39, 8325-8332.

[22] D.A. Johnson, P. Akamine, E. Radzio-Andzelm, N. Madhusudan, S.S. Taylor, Dynamics of cAMP-dependent protein kinase, *Chem. Rev.*, (2001), 101, 2243-2270.

[23] I. Bartova, M. Otyepka, Z. Kriz, J. Koca, Activation and inhibition of cyclin-dependent kinase-2 by phosphorylation; a molecular dynamics study reveals the functional importance of the glycine-rich loop, *Protein Science*, (2004), 13, 1449-1457.

[24] J.K. Halmes, M.J Solomon, The role of Thr160 phosphorylation of CDK2 in substrate recognition, *Eur. J. Biochem.*, (2001), 268, 4647-4652.

[25] A. Cook, E.D. Lowe, E.D. Chrysina, V.T. Skamnaki, N.G. Oikonomakos, L.N. Johnson, Structural studies on phospho-CDK2/Cyclin A bound to nitrate, a transition state analogue: Implications for the protein kinase mechanism, *Biochemistry*, (2002), 41, 7301-7311.

In: Molecular Simulation Studies in Material ... ISBN 1-59454-607-X
Editor: Kholmirzo Kholmurodov, pp. 173-182

Chapter 14

PROTEIN-LIKE BEHAVIOR OF A FLEXIBLE CHAIN POLYETHYLENE: MOLECULAR DYNAMIC SIMULATIONS WITH WATER

D. A. Kretov and Kh. T. Kholmurodov*
Department of Radiation and Radiobiological Research
Joint Institute for Nuclear Research
141980, Dubna, Moscow region, Russia

Abstract

In the present paper we used molecular dynamics (MD) simulations to study the density and the temperature behavior of a flexible polyethylene (PE) subjected to various heating conditions and to investigate the PE chain conformational changes in a water solvent. First, we have considered the influence of the heating process on the final state of the polymeric system and the sensitivity of its thermodynamic characteristics (density, energy, etc.) for different heating regimes. For this purpose the three different simulations were performed: fast, moderate, and slow heating. Second, we have investigated the PE chain conformational dynamics in water solvent for various simulation conditions and various configurations of the environment. From the obtained results we have got the pictures of the PE dynamical motions in water. We have observed a protein-like behavior of the chain PE, like that of the DNA and the proteins in water, and have also estimated the rates of the conformational changes. For the MD simulations we used the optimized general-purpose DL_POLY code (B. Lustig, Nucleic Acids Research, 1995; G. Lancelot, Proc. Nat. Acad. Sci., 1997) and the generic DREIDING force field. The MD simulations were performed on the parallel computers and special-purpose MDGRAPE-2 machine.

1. Introduction

Flexible protein molecules are well known to roll up in water to take more structured forms or optimal conformations [1]-[3]. A protein or an arbitrary polypeptide chain due to the hydrophobic interactions with water turns into a secondary or even a tertary structure,

*E-mail address: kretov_da@mail.ru, mirzo@jinr.ru

therefore making them possible to perform their specific and unique biological functions. Thus, the hydrophobic interactions, as results of the hydrogen bonding of the corresponding groups or side chains and water, play a central role in the structural conformations and functions of biological structures [1],[3]-[5]. From the experimental point of view the measurements on the protein conformation is an extremely difficult and an ambiguous task [6]-[9]. Moreover, experimental data does not provide a unique answer on a valid structural behavior of proteins. In the one hand, the protein structure maybe well measured in the form of the crystal only (or as a dense solvent). On the other hand, defined in the state of crystal the protein structure may essentially deviate from that one exists under the physiological conditions and temperatures. It turns out that the conformations of the biological objects (structures of DNA, RNA and proteins) in water are most suitable to simulate through the computer molecular modeling or design of the synthetic copolymers. There has been a recent trend in nanostructural simulations to make it possible the observation of a protein-like behavior for synthetic copolymers (with the inclusion of water and other ionic solvents) [2],[10],[11]. In the computer molecular simulations such protein-like copolymers with desired properties are easy to generate, thereby playing a huge number of the configurations efficiently. On the other hand, for the polymeric molecules the nature of the dynamical changes in water is similar to the hydrophobic interactions of DNA or proteins [2],[4],[8]-[12]. Thus, simulations on the flexible polymeric chains may provide a useful insight into the mechanisms of the protein-water dynamics or related phenomena. Synthetic copolymers have to collapse, for example, into a nanostructed globules with well-defined hydrophobic cores wrapped in hydrophilic envelopes [2],[11]-[13]. The behavior of the polymeric chains (say, in water) makes possible for a direct exhibition of the hydrophobic (or hydrophilic) parts. So, simulation on the polymer and water structural dynamics is a valuably useful stage in the understanding of proteins functions and their conformational changes. Synthetic copolymers with a protein-like property cause a special interest in material sciences due to the modern innovations in nano- and biopolymer technology. The detailed structural pictures of polymers, based on their microscopic transition behavior, represent an interest for the industry of polymer materials, for the material modification and technological aspects. It is worth noting computer MD simulation method is one of the most convenient techniques in studying the polymer phase transitions and other technologically interesting phenomena (structural conformations, chain-folding processes, time and temperature behavior, etc.) [10]-[17]. The MD simulation allows us to estimate the density-temperature changes of polymeric chains, to study their time dependent thermodynamic properties in solvents and explore nontrivial polymer-surface interactions [10],[18],[21]. In the present work we were aimed to investigate via MD simulations the density and temperature behaviors of a flexible polyethylene (PE) and to investigate for a PE chain a protein-like behavior in an explicit (water) solvent. The MD simulations have been performed on the basis of the optimized general-purposes DL_POLY code [20],[21] and generic DREIDING force field for molecular simulations [10].

2. Details of Simulations

The MD simulations of a single flexible polyethylene (PE) chain of about 500 (mixed $CH_2 +$ CH_3) units in length (each of CH_2 or CH_3 group is treated as a united atom) were performed using DREIDING force field. A snapshot of the initial configuration of the PE chain used in our simulations is shown in Fig.1.

Figure 1. Initial configuration of a polyethylene chain

The initial PE configuration was constructed with the help of the built-in Java interface provided by the DL_POLY package, that supports a number of the utilities for the analyzing of the results and visualization purposes.

The configuration energy of the PE model is a sum of the intramolecular valence (bonded) E_{val} and nonbonded E_{nb} interaction energies. The intramolecular energy E_{val} includes bond stretching, bond angle bending, and torsion and inversion interactions. Nonbonded interactions consist of Van der Waals (VdW), electrostatic and hydrogen bonds. The VdW interactions were calculated by the pairwise 6-12 Lennard-Jones (LJ) potential. The LJ-parameters of the PE model were: C-C - $\varepsilon = 0,0951$, $\sigma = 3,4730$; C-H - $\varepsilon = 0,0380$, $\sigma = 3,1597$; H-H - $\varepsilon = 0,0152$, $\sigma = 2,8464$ (ε, *kcal/mol*; σ, *Å*). We employed the NPT ensemble; a constant temperature of was attained with the Nose-Hoover method; VdW forces cutoff was 10,5Å and Ewald precision parameterα was set to 10^{-4}. The equations of motion were solved by the leapfrog Verlet algorithm with a time step choosen from the interval of $[0,00040,001]ps$. The relaxation constant for the heat bath variable was $0,1ps$.

3. Results and Discussion

3.1. A Flexible PE Chain Subjected to Various Heating Conditions

We have first considered the influence of different heating processes on the final state of the chain PE under the periodic boundary conditions. In Fig.2 the dependence of the temperature (T) vs time (t) is shown for three heating regimes: left picture displays fast; iddle picture -. moderate; and right picture - slow heating. Starting from the same initial state at T(0)=10 K during the system equilibration we have reached a final state of T(t)=300 K; the temperature step (ΔT) was choosen for the fast heating mode 50 K, for the moderate - 50 K, for the slow - 10 K. The starting time t=4 ps coincides with the end of the equilibration process.

For the above three heating processes we have estimated the evolutions of the atomic density of the system. In Fig.3 the behaviors of the particle density are shown for the same three regimes as displayed in Fig. 2 (from left to right).

The resultant PE configurations for the above three heating modes are presented in Figs.4 (from left to right).

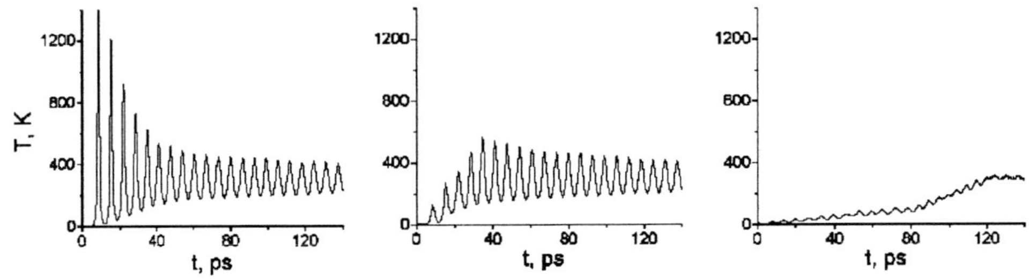

Figure 2. The evolutions of the temperature of the system. The fast (left), moderate (middle) and slow (right) heating regimes are shown

Summarizing the results presented above, we observe that the evolution of the temperature has a clearly different behavior for these three heating regimes. During the fast and moderate heating we observe large temperature changes. At the same time, the slow heating yields satisfactory small temperature T(t) oscillations. In other words, the slow heating mode "smoothly" forms the equilibrium state, which manifests itself as an important stage for the final dynamics.

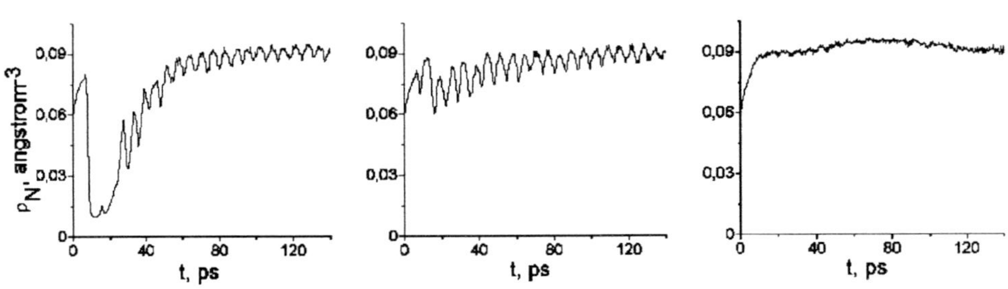

Figure 3. The evolutions of the atomic density in the fast (left), moderate (middle) and slow (right) heating regimes

A final state, achieved at the point t=140 ps, has approximately the same value of T for all heating regimes. For a larger MD time scales ($t > 140ps$) a more significant changes of the temperature values has not been observed, so far the PE system has been analyzed only at the interval from 0 to 140 picoseconds. Next, the evolutions of the atomic density of the system, shown in Figs.3, indicate that the slow heating process leads to an equilibrium density monotonically.

In the fast and moderate heating the visibly high oscillations of the density values are observed. The behavior of the PE system under consideration is sensitive to the dynamics of the heating mode. The high temperature oscillations (the left and the middle pictures in Figs.2-3) result in significant change of the particle number density. However, it is remarkable that the PE chain does not respond to these significant temperature changes.

The resulting particle number densities are the same for all heating modes. In the equilibrium state (t=140 ps) we have have the following values, respectively: $\rho_{fast} = 0,165532$, $\rho_{moderate} = 0,160132$, $\rho_{slow} = 0,165256$ (g/cm^3). Thus, the large temperature oscillations, which occur in a very short period of time (viz. [10-20] ps), do not break the PE chain on constitutive fragments, so that the polymeric chain keeps its stability over a wide range of the temperature oscillations.

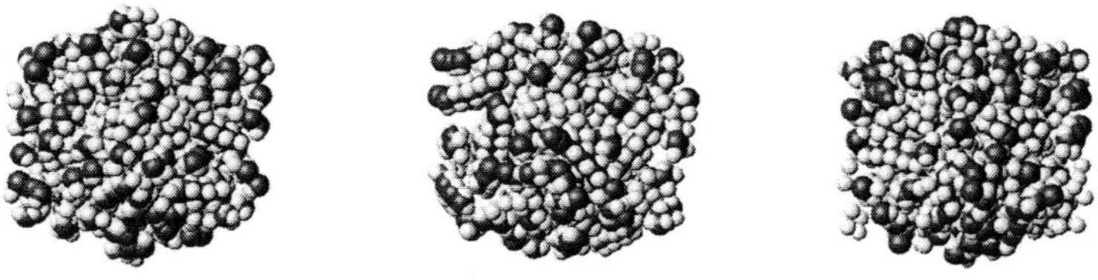

Figure 4. The resultant PE configurations in the fast (left), moderate (middle) and slow (right) heating regimes

3.2. A Protein-Like Behavior of the Flexible PE Chain in Water

Next, we simulated the conformation dynamics of the PE chain in a solvent. For the solvent molecules we have used the TIP4P water in the different configurations and for the different cell geometries. In Fig.5 an example of the initial configuration of the PE chain and water is shown.

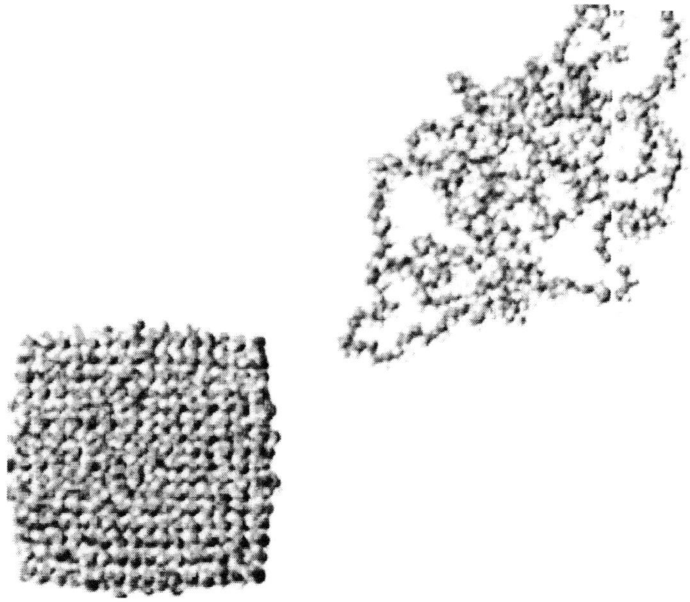

Figure 5. The initial configuration of a polyethylene chain with cubic ice (left bottom)

To reach an equilibration for the PE chain and cubic ice, the system temperature was gradually from 0 to 300 K under the different heating conditions described above.

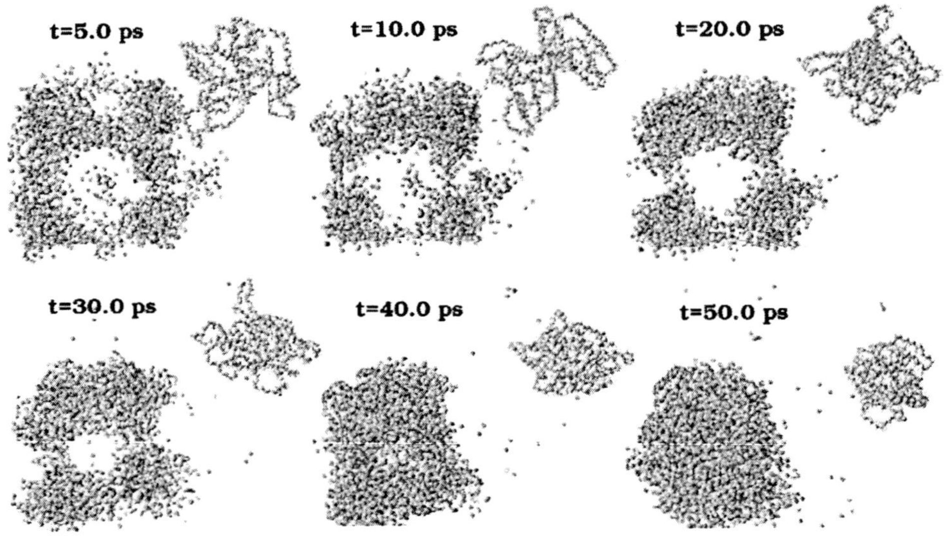

Figure 6. The six sequential snapshots of the PE chain + water configurations are presented. For the water solvent we have employed TIP4P water model at the different configuration and geometry of the initial cell

In Fig.6 the six sequential snapshots of the PE + water at the moments of t=5, 10, 20, 30, 40, 50 ps are displayed. Starting from the well-separated state, the PE chain is brought in contact with water. The PE chain, however, does not mix with the solvent molecules: it rather hides their fragments inside a central region ("a hydrophobic core"). Such behavior is similar to that of a DNA or a protein interacting with water.

The dynamical behavior of the PE chain is affected by the water conformation. The shape of the water configuration becomes a visibly non-symmetric due to the PE-water interactions.

In Fig.6 a cavern-like formation is seen as well: the size of the latter correlates with a water-PE contact distance. These two parameters, the cavern diameter and the PE-water distance, are drawn in Fig.7.

The calculated values of the D_{cav} (water cavern diameter) and D_{min} (PE-water contact distance) are presented in Figs.8. Starting from the formation stage at t=0 the diameter of the water cavern (left picture) grows up to the point of $t \sim 20ps$. That temporal point is also correlates with a PE-water distance to reach of a maximum value: within the same period we observe the increase of the PE-water contact distance.

Starting from the instant of about $t \sim 20ps$ the water cavern closes gradually; also starting from this point the PE-water contact distance become constant. In other words, the PE chain folds into a preferable state which is similar to the protein conformation in water solution.

3.3. The Dynamics of a Cavern-Like Formation

To make comparison with the behaviour of a cavern-like formation, presented in Figs.5-8, we have investigated the dynamical changes of water without the PE chain at the same simulation conditions and the cell size as considered above.

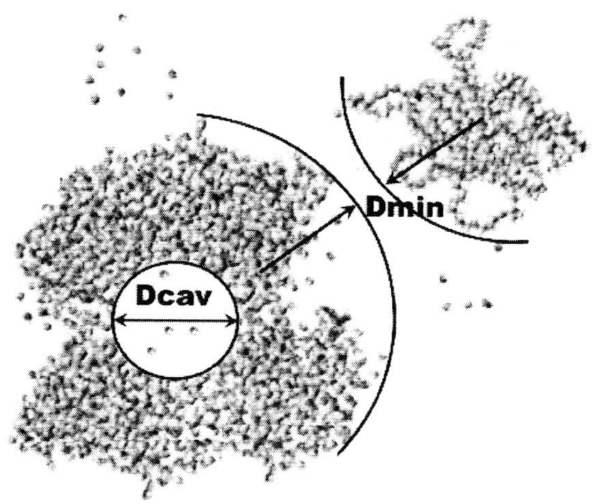

Figure 7. The distance correlation PE chain + water configurations are presented through the two distances: diameter of the water cavern and PE-water contact distance

Briefly, we have fulfilled one more set of calculations just excluding the PE chain from the system. In Fig.9 the initial configuration of water, a cubic ice, is shown.

In Fig.11 the six sequential snapshots of the water solvent are displayed. If to compare these pictures with those in Fig.6, it is easy to see that the water configurations are almost spherically shaped. It is also worth noting that in the absence of the PE chain the changes in size of the caver-like formation can be distinguished well. Thus, the influence of the flexible PE chain on the cavern configuration (or vice versa) we estimated via the changes of a cavern diameter.

In Fig.10 the water cavern diameter is shown for the same cell and distant positions as in Fig.7.

The calculated values of the cavern diameter are presented through the Fig.12.

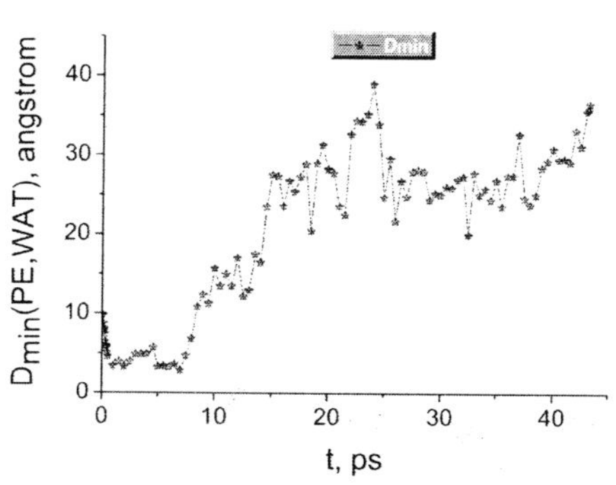

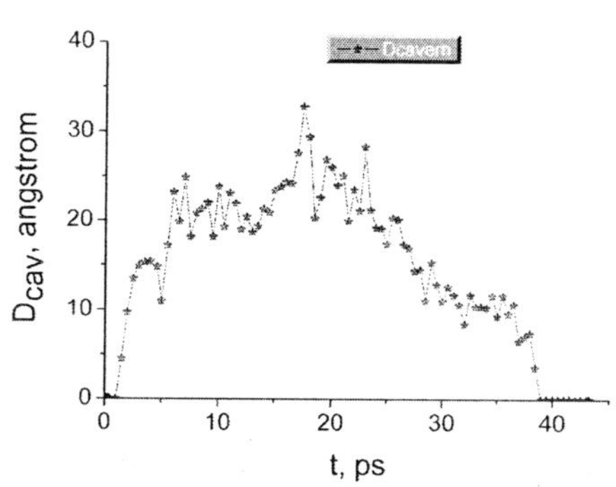

Figure 8. The two distances vs time for the PE chain + water configuration as shown in Fig.7 are presented: left - diameter of the water cavern, right the PE-water contact distance

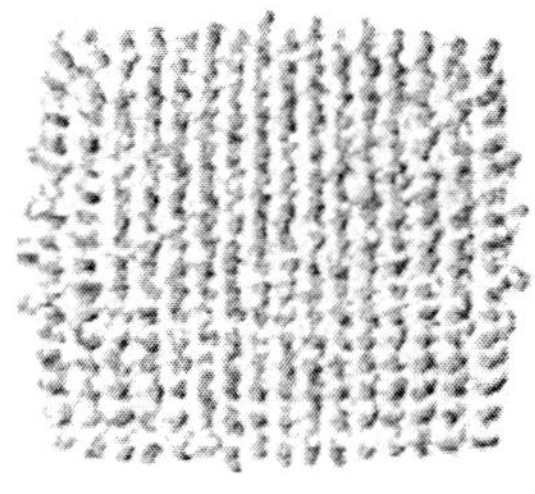

Figure 9. The initial configuration of a polyethylene chain with cubic ice (left bottom)

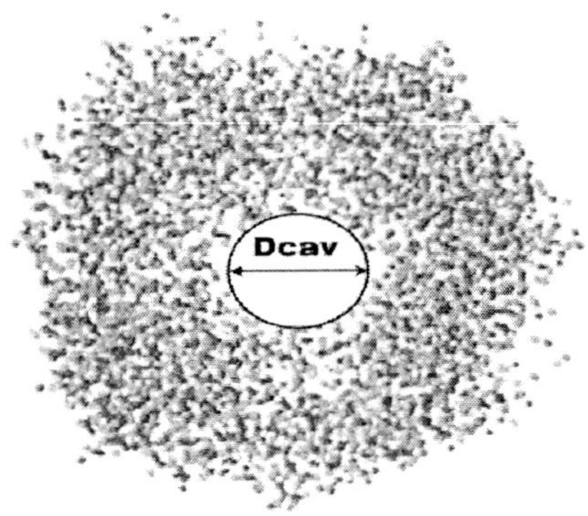

Figure 10. The water cavern diameter in the same cell-size as in Fig.7

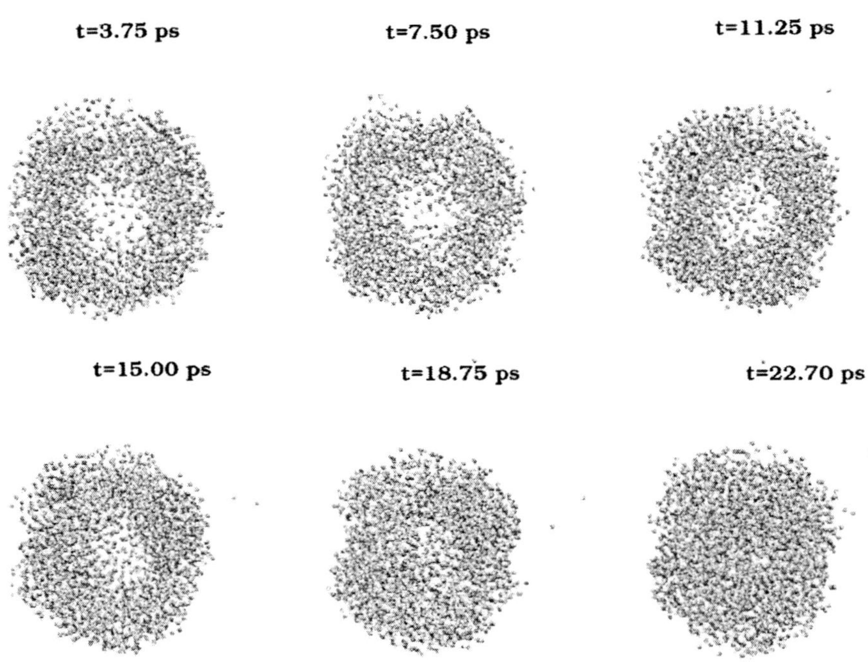

Figure 11. The six sequential snapshots of the water configurations are presented. For the water solvent we have employed TIP4P water model at the same initial configuration and geometry of the cell, as in Fig.6

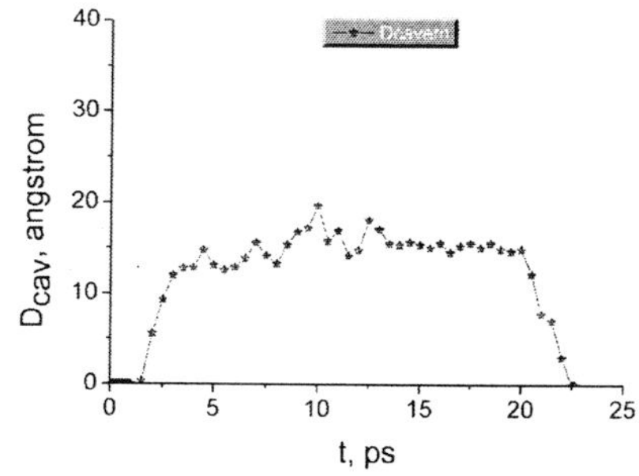

Figure 12. The diameter of the water cavern vs time

From the time dependences for the D_{cav} (viz. Fig.12 and Fig.8) it can be easily observed that the values of the cavern diameter are possesses the visibly different behaviours. The most interesting point is that the D_{cav} in the absence of the PE chain keeps around of some average value (15 Å). In other words, the water cavern goes through the long oscillations before to close abruptly. Also the values of D_{cav}, as seen from Fig.12, have to be approximately two times smaller in comparison with Fig.8. Thus, the presence of the PE chain essentially changes the dynamics of the water cavern and "modifies" the cavern diameter. The process of the cavern-like formation under almost the same thermodynamical and cell-geometry conditions without the PE chain has to be characterized by a fast dynamics. It is remarkable that the influence of the PE chain on the water dynamics is accompanied by its protein-like hydrophobic behaviour, which should be taken into account for the studies of the flexible biomolecular systems involving solvent.

References

[1] V.M. Stepanov, Molecular biology. Structure and functions of proteins, Moscow, Visshaya shkola., 1996 (*in russian*).

[2] M. Karplus, G.A. Petsko, Molecular dynamics simulations in biology, *Nature*, (1990), 347, 631-639.

[3] J.C. Smith, "Protein Dynamics and Hydration: Scattering meets Computer Simulation", *In: "Books of Abstracts" of the Conference "Physics meets Biology-Scattering and Computer Simulationsin-"*, Donostia-San Sebastian, June 26-28, 2003.

[4] M.-C. Belissent-Funel, J.-M. Zanotti, and S.H. Chen, Slow dynamics of water molecules on surface of a globular protein, *Faraday Discussion*, (1996), 103, 281.

[5] S. Dellerue and M.-C. Belissent-Funel, Relaxation dynamics of water molecules at protein surface, *Chem. Phys.*, (2000), 258, 315-325.

[6] K. Wthrich, NMR of Proteins and Nucleic Acids, John Wiley & Sons, 1986.

[7] M. Sattler, S.W. Fesik, Use of deuterium labelling in NMR: overcoming a sizeable problem, *Structure*, (1996), 4, 1245-1249.

[8] B. Lustig, Consistencies of individual DNA base-amino acid interactions in structures and sequences, *Nucleic Acids Research*, (1995), 23(22), 4707-4711.

[9] G. Lancelot, Selective recognition of nucleic acids by proteins: the specificity of guanine interaction with carboxylate ions, *Proc. Nat. Acad. Sci.*, (1997), 74(11), 4872-4875

[10] S.L. Mayo, B.D. Olafson, and W.A. Goddard III, *J. Phys. Chem.*, (1990), 94, 8897.

[11] A.R. Khokhlov, "Biomimetic Sequence Design and Evolution of Sequences in Copolymers", *In: "Books of Abstracts" "Physics meets Biology -Scattering and Computer Simulationsin-"*, Donostia-San Sebastian, June 26-28, 2003.

[12] J.-M. Zanotti, M.-C. Belissent-Funel, and J. Parello, "Hydration-coupled dynamcis in proteins studied by neutron scattering and NMR. The case of the typical EF-hand calcium-binding parvalbumin", *Biophys. J.*, (1999), 76, 2390.

[13] S. Dellerue, A.J. Petrescu, J.C. Smith, and M.-C. Belissent-Funel, "Radially softening diffuse motions in a globular protein", *Biophys. J.*, (2001), 81, 1666-1676.

[14] N.L. Allinger, *J. Comput. Chem.*, (1987), 8, 581.

[15] S.M. King and T. Cosgrove, *Macromolecules*, (1993), 26, 5414.

[16] J. Budzien, C. Raphael, M.D. Ediger, and J.J. de Pablo, *J. Chem. Phys.*, (2002), 116, 8209.

[17] M. Fukuda and H. Kikuchi, *J. Chem. Phys.*, (2000), 113, 4433.

[18] H.X. Guo, X.Z. Yang, and T. Li, *Phys. Rev. E*, (2000), 61, 4185.

[19] T.K. Xia and U. Landman, *Science*, (1993), 261, 1310.

[20] P.V. Krishna Pant, J. Han, G.D. Smith, and R.H. Boyd, *J. Chem. Phys.*, (1993), 99, 597.

[21] W. Smith, T.R. Forester, *Molecular Graphics*, (1993), 14, 136.

[22] Kh. Kholmurodov, et al., *Comput. Phys. Commun.*, (2000), 125, 167.

In: Molecular Simulation Studies in Material ... ISBN 1-59454-607-X
Editor: Kholmirzo Kholmurodov, pp. 183-187

Chapter 15

SIMULATION OF NONLINEAR PROCESS OF DISTRIBUTION OF ELASTIC PERTURBATIONS IN THE MOLECULAR CHAIN OF POLYMERIC CRYSTALLITE

*E. Dushanov*and U. Gafurov*
Institute of Nuclear Physics UzAS
702132 Tashkent, Ulugbek, Uzbekistan

Abstract

Simulation of distribution of elastic perturbation along a linear polyethylene chain in polymeric crystallite was carried out. Like that in Frenkel-Contor model influence of crystal substrate was taken into account by introduction of sine intermolecular potential. For the description of interatomic interaction along the molecular chain the Morse potential was used. In the system of $N = 100$ methylene groups with and without structural defects in a chain, oxidized groupings CO with a variation of their number (1 and 2 groupings), the place of localization and interposition were considered. The study was carried out by the method of molecular dynamics (MMD). Elastic perturbation was transferred along molecular chain by a stretching of trailer bond, the dynamics of which was traced in a time interval. Abnormal character of this perturbation in places of localization of defects were shown.

1. Introduction

Problems of distribution of perturbation dynamics and energy localization in crystal structures, in particular, in molecular and polymeric connections, is of significant interest and receives a lot of attention. For example, these problems for linear molecular chains were mentioned in [1]-[6]. Real macromolecular polymeric bonds have various structures, chemical defects and conformations [7]-[12]. They are therefore of special interest for our consideration. This work is devoted to the simulating of influence of chemical structural defects, such as oxidized $C = O$ groupings in a polymeric chain, on the process of distribution of elastic perturbations.

*E-mail address: dushanov@inp.uz

2. Molecular Model and Research Method

The chain of 100 beads with 2 groupings on a crystal substrate and the same chain containing $C = O$ groupings with a variation in their number (1 or 2), in places of their localization in a linear polyethylene chain and interpositions were considered. The Morse potential was used as interatomic interaction potential

$$U(r) = D(\exp(-\alpha(r - r_0)) - 1)^2, \tag{1}$$

where D is the bond dissociation energy, r_0 is the distance between units. The influence of the crystal substrate was taken into account by introduction of the sine potential, similar Frenkel-Contors in model [11],[13]:

$$U(x) = U_0(1 - \cos\frac{2\pi}{a_0}x), \tag{2}$$

where U_0 is height of intermolecular potential, a_0 is the period (equilibrium distance between units) a polymeric crystal lattice.

The equation of motion of atoms in the initial configuration has the form

$$m_i\ddot{x}_i = -\frac{2\pi}{a_0}U_0 \sin\frac{2\pi}{a_0}x_i + 2D\alpha \sum_{j=1}^{N} \frac{r_{ij}}{|r_{ij}|} \exp(-\alpha(r_{ij} - r_0))(\exp(-\alpha(r_{ij} - r_0)) - 1), \tag{3}$$

where x_i is i-th unit displacement, $r_{ij} = x_i - x_j + a_0(i - j)$.

Oxidized $C = O$ groupings were built in various places of the polyethylene chains

$$\begin{array}{c} — \mathrm{CH2} — \cdots — \mathrm{CH2} — \mathrm{CH2} — \mathrm{CH2} — \cdots — \mathrm{CH2} — \\ — \mathrm{CH2} — \cdots — \mathrm{CH2} — \mathrm{CO} — \mathrm{CH2} — \cdots — \mathrm{CH2} — \end{array}$$

The study of distribution of perturbations were carried out by the method of molecular dynamics. Elastic perturbation was transferred along the molecular chain by a stretching of trailer bond on 10% from its length. The distribution of deformations and kinetic energy along a molecular chain was studied with the help of computer modeling.

Dynamics of distribution of perturbation was traced during 0.3 ps by numerical integration of the equation of motion (3). The numerical Verlet algorithm [14] with time step $\delta t = 0.1 \cdot 10^{-3}$ ps was used.

The following values of the parameters were used: $D = 3.3$ eV (73 kcal/mol), $r_0 = 1.585$ Å, $a_0 = 1.54$ Å. For methylene groupings size U_0 was set equal 0.57 kcal/mol (0.025 eV), and for C=O groupings - 1.15 kcal/mol (0.05 eV).

3. Results and their Discussion

In fig. 1 the results of simulation of the dynamics of distribution of deformation of interatomic bonds along a linear faultless chain in various time intervals from beginning of simulations ($1.2 \cdot 10^{-3}$ ps, $33.3 \cdot 10^{-3}$ ps, $7 \cdot 10^{-2}$ ps, $11.1 \cdot 10^{-2}$ ps, $15 \cdot 10^{-2}$ ps) are presented.

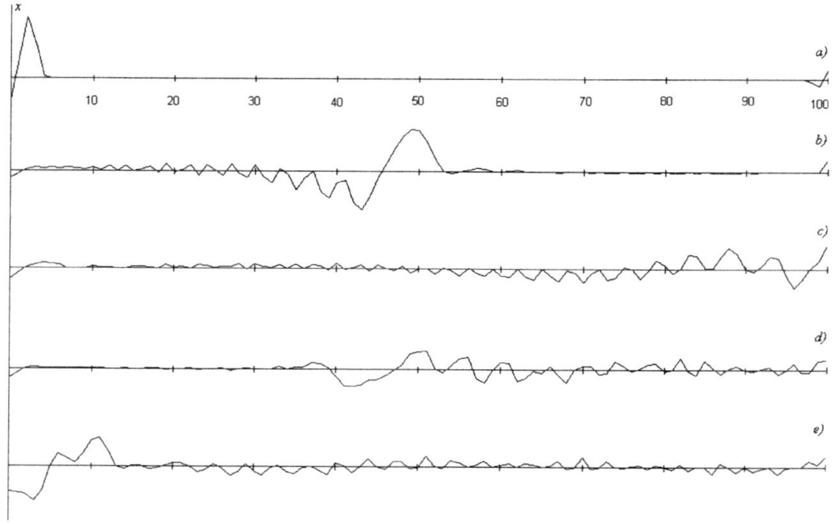

Figure 1. Dynamics of distribution of deformations of interatomic bonds along a linear chain in various time intervals: a) $t = 1.2 \cdot 10^{-3}$ ps; b) $t = 33.3 \cdot 10^{-3}$ ps; c) $t = 7 \cdot 10^{-2}$ ps; d) $t = 11.1 \cdot 10^{-2}$ ps; e) $t = 15 \cdot 10^{-2}$ ps

As it is seen from the figures, the chain perturbation is distributed along the chain up to the end of any anomalies.

In fig. 2 the results of simulations of dynamics of the distribution of deformation of interatomic bonds along a linear chain with one oxidized C=O a grouping in the middle of a chain (the 50th unit) on the expirations of various time intervals from the beginning of perturbation for a chain are shown.

At the initial stage the character of distribution of perturbation is actually identical to that of a linear faultless chain, with the considered defect (fig. 2). As the perturbation reaches the position of of defect localization, part of its kinetic energy is reflected from defect (fig. 2b).

Total kinetic energy for a chain with the defect at the time instant ($t = 33.3 \cdot 10^{-3}$ ps) was about 0.34 eV, while for the faultless chain it is 0.33 eV. This effect is repeated after return of the reflected perturbation from the end of a molecular chain (see, fig. 2d).

Similar results have been yielded for molecular chains with two C=O groupings. In fig. 3 the results of the performed calculations for the dynamics of distribution of deformation of interatomic bonds for a molecular chain with C=O groupings in 33 and 70-th units are presented.

As it is easily seen, perturbation for such structural configuration fades eventually. It was possible to notice abnormal growth and the delay of kinetic energy in the positions of defects. For a configuration of molecular chains shown in fig. 3d and 3e the values of kinetic energy are $E_k = 0.47$ eV and $E_k = 0.59$ eV, respectively.

Acknowledgments

This work is supported by STCU Fund (Project №3009).

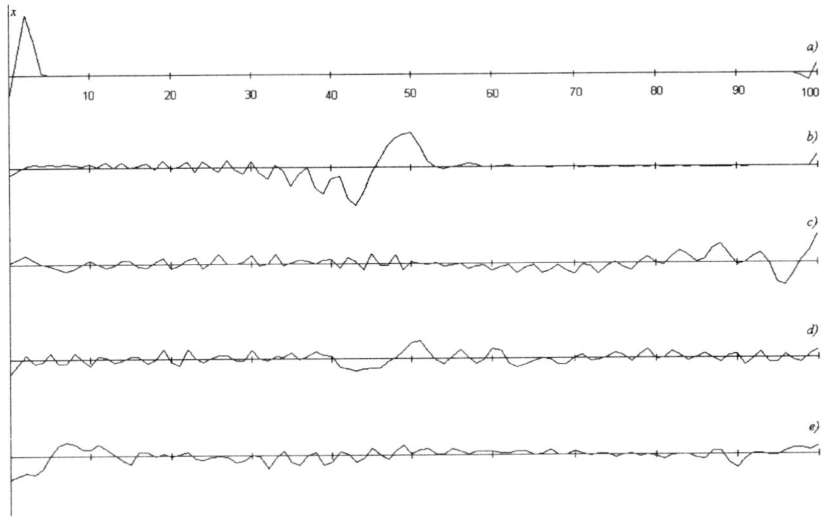

Figure 2. Dynamics of distribution of deformation of interatomic bonds of a linear chain with C=O a grouping on the expirations of various time intervals: a) $t = 1.2 \cdot 10^{-3}$ ps; b) $t = 33.3 \cdot 10^{-3}$ ps; c) $t = 7 \cdot 10^{-2}$ ps; d) $t = 11.1 \cdot 10^{-2}$ ps; e) $t = 15 \cdot 10^{-2}$ ps

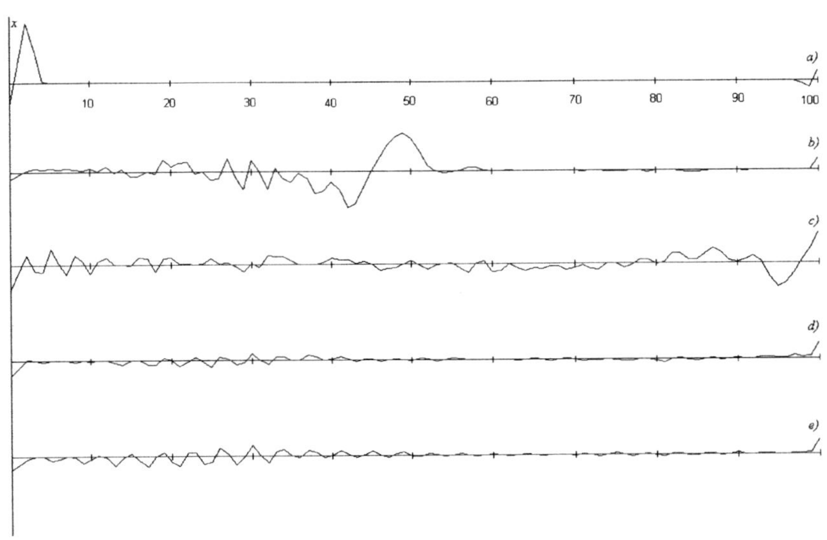

Figure 3. Dynamics of distribution of deformation of interatomic bonds along a chain. Shown for different time intervals: a) $t = 1.2 \cdot 10^{-3}$ ps; b) $t = 33.3 \cdot 10^{-3}$ ps; c) $t = 7 \cdot 10^{-2}$ ps; d) $t = 11.1 \cdot 10^{-2}$ ps; e) $t = 15 \cdot 10^{-2}$ ps for a chain with two defects – the oxidized groupings in 33 and 70-th units

References

[1] H. Graener and F. Laubreau, *Chem. Phys. Lett.*, (1987), 133, 378.

[2] J.R. Reimers, *Chem. Phys.*, (1990), 146, 89.

[3] D.C. Doherty and A.J. Hopkinger, *Phys. Rev. Lett.*, (1994), 72, 661.

[4] U. Gafurov, *Proc. SPIE*, (1999), 3687, 201-204.

[5] H.L. Chen, C.L. Lee and C.L. Chen, *Polymer*, (1994), 35(72).

[6] N. Balabaev, *Phys. Rev. E*, (2001), 64, 036702 (1-8).

[7] S.V. Dmitriev, M. Itoh, M. Shinohara, *Progress of Theor. Suppl*, (2000), 138, 574.

[8] D.H. Reneker, *Polymer*, (1994), 35(8), 1909.

[9] G.R. Strobl, *The Physics of Polymers*, 2nd ed., Springer, Berlin, 1999.

[10] S.T. Pantelides, *Phys. Today*, September, (1992), 67.

[11] Questions of the theory of defects in crystals, *The collection of proceedings*, Nauka, Leningrad, 1987, 176 (in russian).

[12] A.N. Orlov, Introduction in the theory of defects in crystals, Visshaya shkola, Moscow, 1983, 144 (in russian).

[13] H.H. Kausch and J.G. Plummer, *Polymer*, (1994), 35(18), 3848.

[14] L. Verlet, Computer "experiments" on classical fluids. I. Thermodynamical properties of lennard-jones molecules, *Phys. Rev.*, (1991), B44(5), 2358-2361.

INDEX

A

B

E

F

G

H

I

J

K

L

M

N

O

P

Q

R

S

T

U

V

W

X

Y